# Fluidization and Fluid Particle Systems: Fundamentals and Applications

## *Liang-Shih Fan, editor*

Bavarian, F.
Beaton, W.I.
Bergougnou, M.A.
Bradshaw, W.M.
Briens, C.L.
Buttke, R.D.
Cai, P.
Chan, I.
Chen, S.P.
Chen, Y-M
Dhodapkar, Shrikant
DiFelice, R.
Fan, L.-S.
Findlay, J.
Foscolo, P.U.
Geldhart, Derek
Gianfreda, L.
Gibilaro, L.G.
Greco, G. Jr.
Ho, T.C.
Huang, C.C.
Jin, Y.
Kiel, J.H.A.

Klinzing, George E.
Knowlton, T.M.
Kono, H.O.
Krishnan, R.P.
Li, K.Y.
Massimilla, L.
Mohrman, G.B.
Myler, Craig A.
O'Brien, T.J.
Peng, S.H.
Pirozzi, D.
Pratsinis, Sotiris E.
Rapagna, S.
Shaffer, F.D.
Sishtla, C.
Syamlal, M.
van Swaaji, W.P.M.
Vogiatzis, A.L.
Wang, Z.W.
Xi, M.
Young, J.M.
Yu, Z.Q.
Zaitash, Abdolreza

**AIChE Symposium Series**

1989

*Published by*
**American Institute of Chemical Engineers**

Copyright 1989

American Institute of Chemical Engineers
345 East 47 Street, New York, N.Y. 10017

*AIChE shall not be responsible for statements or opinions advanced in papers or printed in its publications.*

Fluidization and fluid particle systems : fundamentals and applications / Liang-Shih Fan, editor.
  p. cm. — (AIChE symposium series ; no. 270, v. 85 (1989))
  Papers presented at the AIChE Annual Meeting in Washington, D.C., Nov. 27–Dec. 2, 1988.
  ISBN 0-8169-0468-5

  1. Fluidization—Congresses.   I. Fan, Liang-Shih.
II. American Institute of Chemical Engineers.   Meeting (1988 : Washington, D.C.)   III. Series: AIChE symposium series : no. 270.
TP156.F65F5734 1989
660'.284292—dc20                                        89-17725
                                                         CIP

Authorization to photocopy items for internal or personal use, or the internal or personal use of specific clients, is granted by AIChE for libraries and other users registered with the Copyright Clearance Center (CCC) Transactional Reporting Service, provided that the $2.00 fee per copy is paid directly to CCC, 21 Congress St., Salem, MA 01970. This consent does not extend to copying for general distribution, for advertising, or promotional purposes, for inclusion in a publication, or for resale.

Articles published before 1978 are subject to the same copyright conditions and the fee is $2.00 for each article. AIChE Symposium Series fee codee: 0065-8812/89 $2.00.

# FOREWORD

This volume of the AIChE Symposium Series encompasses thirteen papers selected from those presented in the four sessions at the AIChE Annual Meeting in Washington, D.C., November 27 through December 2, 1988. This meeting was highlighted with a theme which addressed the emerging technology in a changing environment and its research needs. In line with the theme, the Fluidization and Fluid-Particle Program planned a session on "Novel Fluidization and Fluid-Particle Systems" which emphasized the role of advanced materials and biotechnology in engineering. In fluidization systems, the former is predominently associated with the fine powder processing, while the latter is with the liquid or gas-liquid processing with bioparticles.

The papers presented in this volume cover both the fundamentals and applications in these two emerging areas. For example, in advanced materials, papers include flowability of fine powder and silane-powder decomposition reaction. In biotechnology, papers include mixing in a circulating liquid fluidized bed and development of a circulating fluidized bed fermentor. In addition, papers of traditional importance are incorporated. Examples are pneumatic conveying, flow regime transition, fines elutriation, computer simulation and flue gas cleaning. The contents of the volume are grouped into two parts: Fundamentals and Applications.

I would like to acknowledge the outstanding work of the chairmen and co-chairmen of the sessions for their planning and paper selection, and the reviewers who provided comments for the publication of this volume.

Liang-Shih Fan, *editor*
Department of Chemical Engineering
The Ohio State University

# CONTENTS

FOREWORD .................................................................................. III

## Part I—Fundamentals

ACCELERATION ZONE STUDIES IN PNEUMATIC CONVEYING SYSTEMS AT VARIOUS INCLINATIONS
................ Shirikant Dhodapkar, Abdolreza Zaltash, Craig A. Myler and George E. Klinzing  1

A THEORETICAL MODEL FOR THE HYDRODYNAMICS OF GAS-SOLID TRICKLE FLOW
OVER REGULARLY STACKED PACKINGS ................................. J.H.A. Kiel and W.P.M. van Swaaij  11

COMPUTER SIMULATION OF BUBBLES IN A FLUIDIZED BED ....................... M. Syamlal and T.J. O'Brien  22

PARTICLE MIXING IN A CIRCULATING LIQUID FLUIDIZED BED .... R. Di Felice, L.G. Gibilaro, S. Rapagna and P.U. Foscolo  32

EFFECT OF OPERATING TEMPERATURE AND PRESSURE ON THE TRANSITION
FROM BUBBLING TO TURBULENT FLUIDIZATION ................ P. Cai, S.P. Chen, Y. Jin, Z.Q. Yu and Z.W. Wang  37

THE EFFECT OF FLOW CONDITIONERS ON THE TENSILE STRENGTH OF
COHESIVE POWDER STRUCTURES ........................... H.O. Kono, C.C. Huang, M. Xi and F.D. Shaffer  44

TRANSIENTS IN BED EXPANSION OF A THREE-PHASE FLUIDIZED BED
............................... Y.-M. Chen, F. Bavarian, L.-S. Fan, R.D. Buttke and W.I. Beaton  49

## Part II—Applications

AN OVERVIEW OF MATERIAL SYNTHESIS BY AEROSOL PROCESSES ............................. S.E. Pratsinis  57

SELECTED APPLICATIONS OF ULTRA-RAPID FLUIDIZED (URF) REACTORS:
ULTRAPYROLYSIS OF HEAVY OILS AND ULTRA-RAPID CATALYTIC CRACKING
............................... A.L. Vogiatzis, C.L. Briens and M.A. Bergougnou  69

PREDICTION OF SILICON-POWDER ELUTRIATION IN A FLUIDIZED-BED
REACTOR FOR THE SILANE DECOMPOSITION REACTION ....................... K.Y. Li, S.H. Peng and T.C. Ho  77

THE EFFECT OF SYSTEM PARAMETERS ON FINES GENERATION IN FLUIDIZED
LIMESTONE/COAL-CHAR MIXTURES ........................ C. Sishtla, I. Chan, J. Findlay and T.M. Knowlton  83

LABORATORY TESTING OF A FLUIDIZED-BED DRY-SCRUBBING PROCESS FOR
THE REMOVAL OF ACIDIC GASES FROM A SIMULATED INCINERATOR FLUE GAS
....................... W.M. Bradshaw, R.P. Krishnan, J.M. Young and G.B. Mohrman  94

DEVELOPMENT OF A CIRCULATING FLUIDIZED BED FERMENTOR:
THE HYDRODYNAMIC MODEL FOR THE SYSTEM ......... D. Pirozzi, L. Gianfreda, G. Greco Jr. and L. Massimilla  101

CHALLENGES IN FLUIDIZED BED TECHNOLOGY ....................................... Derek Geldart  111

INDEX ..................................................................................... 122

# ACCELERATION ZONE STUDIES IN PNEUMATIC CONVEYING SYSTEMS AT VARIOUS INCLINATIONS

Shrikant Dhodapkar, Abdolreza Zaltash,
Craig A. Myler and George E. Klinzing ■ Department of Chemical and Petroleum Engineering, University of Pittsburgh, Pittsburgh, PA 15261

Studies in the acceleration zone were carried out for 0.0504 m diameter systems at various inclinations. Experiments were carried out with four different particles (67 $\mu$m, 450 $\mu$m, 900 $\mu$m glass and 400 $\mu$m iron ore) for a wide range solids and gas flow rate. The static pressures and the fluctuations were recorded at various locations along the pipe. The extent of the acceleration region was determined by analyzing the pressure profile. The expression for the acceleration length was analytically derived from the force balance. Predictions of the model compare favorably with the experimental data.

## INTRODUCTION

Pneumatic transport has been used as a viable technology for over a century. Despite its diversified technology with numerous applications in various industries, the theoretical analysis for gas-solids flow is far from perfect. With the increased acceptance of pneumatic conveying as a viable technology, a variety of applications have been found to convey solids for short distances. In such systems, the solids are often collected before they reach the steady state conditions. A need was therefore felt to study the behavior of gas-solid systems at the entry section. This region is termed as the acceleration zone because in this region the particles and the gas accelerate from their inlet conditions to the steady state conditions.

Most of the models based on fundamental equations of mass and momentum balance are too complicated for practical applications. However, the basic model by Yang and Keairns [1] and further analysis

Chemical and Petroleum Engineering Department, University of Pittsburgh, Pittsburgh, PA
* presently with ORNL, Oak Ridge, TN
** presently with U. S. Army, Baltimore, MD

by Enick et al. [2] provides a useful set of equations. The effect of voidage variation has been addressed in the latter. Another physical model proposed by Shi and Michaelides [3] has been simplified from the theoretical viewpoint but is useful for calculation of acceleration length.

The earliest attempt reported in the literature on the study of acceleration length was made by Papai [4]. The most quoted expression for the determination of acceleration length in horizontal systems is that given by Rose and Duckworth [5]. In another study, Shimizu et al. [6] suggested a format for correlation of dimensionless form of entry length ($L_a/D$), for upward flow as a function of apparent flow suspension flow Reynolds number ($Re_m$). Enick and Klinzing [2] have correlated a group of three data sets using a format suggested by Shimizu et al. [6].

Basic equations of force balance, with a broad grouping of forces, have been the basis of the Yang's model [7] for pressure drop prediction. Comparison of the results of numerical analysis of various models in vertical pneumatic conveying of single sized and a binary mixture of particles has been presented by Kmiec and Leschonski [8] who concluded that it is possible to interrelate the pressure and velocity distribution along the pipe axis. Recent attempts to model the motion of solid particles in gaseous streams confined by walls (channel or pipe flow) has been done by Michaelides [9] and he concluded that the particles constantly accelerate between collisions and finally reach a "pseudo-steady state".

**EXPERIMENTAL METHOD**

During the course of experiments, four systems were constructed with 0.0504 m (2 inch) pipe at various inclinations. The horizontal system shown in Figure 1 had a total length of about 30 meters. The configuration resembled U shape with two Tee bends. The vertical system was about 6 meters long and consisted of two long radius 90 degrees bends at the end for solids return to the collection system. The inclined systems (30 and 60 degrees) were laid out in such a way as to avoid inclusion of elements of orientation other than those desired. This was necessary in order to avoid the effect of elements of other orientation on the flow in the inclined systems. Copper tubing (I.D. 0.0504 m) was used for the conveying gas-solids mixture. The selection of copper tubing was necessary to eliminate the problems of electrostatic discharge. Solids were introduced into the system using a screw-type feeder. Acquisition of data on particle velocity measurements, pressure drop measurements in acceleration and steady state region, and estimation of gas velocity was achieved by interfacing an IMB-AT with the transducers and electrostatic ring probes.

Experimental Solids

Five different solids were used for experimentation. Their choice was mostly guided by the varied densities, shapes and availability. Table 1 summarizes the various properties.

**Table 1. Characteristics of Solids Used**

| Material | $\rho_p$ (kg/m$^3$) | $D_p$ (μm) | Shape | $U_t$ (m/sec) |
|---|---|---|---|---|
| Glass | 2470 | 67 | Sphere | 0.46 |
| Glass | 2395 | 450 | Sphere | 3.97 |
| Glass | 2464 | 900 | Crushed | 7.45 |
| Iron Oxide | 5004 | 400 | Flaky | 5.87 |

**ANALYSIS**

Estimation of Acceleration Length

The static pressure is measured at various points along the pipe length. The pressure gradient is steep in the initial section and the gradient decreases downstream until it reaches a reasonably constant value. A representative set of results for the longitudinal pressure distribution in the test pipe is shown in Figure 2.

Effect of Orientation. The acceleration length in the horizontal system ranges between 1.2 to 2.5 m (or $L_a/D$ = 24 to 50). The acceleration length in the vertical system was found to be very small, 0.3 to 0.5 m ($L_a/D$ = 6 to 10). The acceleration length for 30 degrees inclination was closer to the predictions for horizontal orientation. The predictions for 60 degrees inclination were smaller than 30 degrees but no definitive conclusions could be made because the acceleration region is believed to lie in the viewing section where no tap could be placed.

An explanation is needed for the large acceleration lengths in horizontal systems. From the various flow patterns observed and recorded pressure profiles, it is possible to conclude that a large amount of energy is required to "uniformly" disperse the solids in the pipe after solids have been fed into the system. This decreases the power (energy) available to accelerate the particles. Hence, it takes longer to establish a steady state. In vertical systems, the solids fed into the system mix uniformly at the feed point due to turbulence. As no extra energy is spent to disperse the solids across the flow cross section, the effective acceleration in vertical orientation is much higher. Another important factor that affects the acceleration length is the friction factor. From regression of data obtained from the steady state region, it was found that the friction factor in the vertical orientation was significantly higher than other orientations resulting in much lower acceleration lengths.

Effect of Gas Velocity. In the horizontal system, the acceleration length showed a maxima for 450 µm for various solids flow rates. At higher solids flow rate, the acceleration length decreases with increase in gas velocity within the limits of experimentation, whereas it increases with gas velocity for lower solids flow rate (Figure 4). At higher solids flow rate, greater gas velocities aid in dispersing the solids (which are mostly segregated towards the bottom of the pipe). Therefore, the acceleration length decreases with increase in gas velocity for larger solids flow rates. The effect of gas velocity on acceleration length for vertical systems has been shown in Figure 5.

Effect of Particle Size. It is very difficult to comment on the effect of particle size on acceleration length because of the difference in shapes of the four different particles.

Pressure Fluctuations in Acceleration Region

The pressure fluctuations in the system have been found to be random and stationary (the average value is independent of time of sampling). The average value is the best estimate of the true value whereas the standard deviation indicates the magnitude of the deviation (fluctuation) from true value.

Effect of Gas Velocity. The effect of gas velocity on the nature of pressure fluctuations has been found to be highly dependent on the characteristics of the particular particle, orientation of the system and distance from the feed point. In the case of 450 µm and 67 µm particles in 2-inch horizontal system, minimum fluctuations were observed for gas velocities corresponding to the minimum pressure drop for the system (Figure 6). However, very heavy particles like iron (400 µm) do not contribute to the fluctuations but just settle down in a layer which ultimately leads to blockage of the line. The behavior of the 30 degrees inclination system was closer to the horizontal system and the 60 degrees inclination system was more akin to the vertical system.

Variation With Distance. In a horizontal system the pressure fluctuations decrease in magnitude with distance from the feed point (Figure 7). The pressure fluctuations are related to the extent of uniformity of dispersion of particles in the pipe. In vertical systems the behavior of fluctuations can be inter-related with the mixing patterns near the feed pont. The magnitude of the fluctuations decrease downstream due to particle-wall and particle-particle friction until it reaches a steady value. In some cases, an increase in fluctuations is observed downstream for higher gas velocities. This can be attributed to the shift of the mixing region downstream due to high gas velocity.

## THEORETICAL MODEL

In order to develop the model, it has been assumed that the friction factor for solids flowing in the gas-solids suspension can be estimated by a Konno and Saito [10] type of correlation. The assumption could be verified by correlating the experimental data using non-linear regression techniques. The data required for this analysis has been taken from Zaltash [11] and Myler [12].

Model I - Acceleration Model for Newton's Regime

The present approach is based on Yang's unified theory and the simplified model suggested by Enick et al [2]. Newton's second law can be applied to an individual particle to develop a dynamic particle model. If a material and force balance is done over the differential element, a general force balance can be written for any inclination as

$$dM_s \cdot \frac{dU_p}{dt} = dF_d - dF_g - dF_f \qquad (1)$$

The distance traveled by the particles can be expressed in the form of the particle velocity and time elapsed. For the differential element dL, substitution of $dL = U_p \cdot dt$ can be made. The integration, however, is not straightforward because the friction factor and the voidage are functions of the particle velocity. Starting with the intergral derived from the force balance,

$$L_a = \int_{U_{p1}}^{U_{p2}} \frac{U_p dU_p}{3/4 C_{DS} \varepsilon^{-4.7} \rho_f \frac{(U_f-U_p)^2}{D_p(\rho_p-\rho_f)} - g\sin(\theta) - \frac{2f_s U_p^2}{D}}$$

For the simplicity of presentation, only the final expression obtained after expanding and rearranging the denominator is given here. Thus,

$$L_a = \int_{U_{p1}}^{U_{p2}} \frac{U_p^2 dU_p}{\alpha U_p^3 + \beta U_p^2 + \gamma U_p + \delta} \quad (2)$$

Integration of the total expression for acceleration length between the limits 0 and $U_{ss}$ using the method of partial fractions gives,

$$L_a = \frac{1}{3\alpha} \{ \ln(\alpha U_p^3 + \beta U_p^2 + \gamma U_p + \delta) \Big|_0^{U_{ss}}$$
$$- \frac{1}{\alpha} \sum_{i=1}^{3} C_i \ln[\frac{(U_{pi} - U_{ss})}{U_{pi}}] \} \quad (3)$$

where $U_{ss}$ is the estimated steady state particle velocity. $C_1, C_2, C_3$ are constants in the numerator of partial fractions. The value of the steady state velocity (assumed to be equal to 98% of the feasible root) is taken as the upper limit of integration. The lower limit of zero is justified as the velocity of the particles at the entrance region in the direction of flow is zero.

Model II - Acceleration Model for Stokes' Regime

Using Stokes' model for drag coefficient, an expression for acceleration length can be derived for Stokes' regime. This intergral can be written in a more concise form as,

$$L_a = \int_{U_{p1}}^{U_{p2}} \frac{U_p^2 dU_p}{\alpha' U_p^2 + \beta' U_p + \gamma'} \quad (4)$$

$$L_a = \frac{1}{\alpha'} U^* - \frac{1}{\alpha'^2}(C_1(\text{Ln}(U^* - U_{pa})$$
$$- \text{Ln}(U^*)) + C_2(\text{Ln}(U^* - U_{pb}) - \text{Ln}(U^*))) \quad (5)$$

where $U^*$ is the estimated steady state particle velocity and the constants $C_1$ and $C_2$ are,

$$C_1 = \frac{\beta' U_{pa} + \gamma'}{U_{pa} - U_{pb}}$$
$$C_2 = \frac{\beta' U_{pb} + \gamma'}{U_{pb} - U_{pa}}$$

Discussion of Predictions by the Two Models

The following two test cases were chosen in such a way that the criteria for the flow regimes is satisfied without drastically differing the system parameters. The summary of base values for calculations is given in Table 2.

**Table 2. Summary of Base Values of Parameters**

| Parameters | Model Stokes' Regime | Model Newton's Regime |
|---|---|---|
| $U_g$ | 15 m/s | 15 m/s |
| $W_s$ | 0.1 kg/s | 0.1 kg/s |
| $\rho_p$ | 2400 kg/m3 | 2400 kg/m3 |
| $\rho_f$ | 1.3 kg/m3 | 1.3 kg/m3 |
| $D$ | 0.0504 m | 0.0504 m |
| $D_p$ | 50 μm | 900 μm |
| $\mu_f$ | 1E-5 kg/(m.sec) | 1E-5 kg/(m.sec) |
| $K$ | 3.3 | 59.97 |

Effect of Various System Parameters. The effect of various system parameters was simulated on the model and the predictions were found to confirm most of the observations in the current study and of other investigators.

Effect of Inclination. As observed in the experimental results, a decrease in acceleration length is observed for inclinations varying from 0 degrees (horizontal) to 90 degrees (vertical). The results have been summarized in Figure 8.

Effect of Gas Reynolds Number. With increase in Gas Reynolds number (Re = $DU_g \rho_f/\mu$) or with increase in superficial gas velocity the acceleration length shows a monotonic increase. Figure 9 shows the effect of gas Reynolds number for various inclinations as predicted by Model I for the flow in Newton's regime. The results from Model II for the flow in Stokes' regime are shown in Figure 10.

**Effect of Ratio of Solids Density and Gas Density.** The final value of acceleration length depends on the interaction between effect of density and slip velocity on the drag force and final limits of integration. The effect of density ratio has been summarized in Figure 11 for the two models.

**Effect of Particle Diameter.** The effect of particle diameter is clearly evident if the dimensionless ratio $L_a/D$ is plotted against $D/D_p$. The acceleration length should be extremely small for small particles. In fact, as the particles reach molecular size, the predicted slip should be negligible. The predictions of Model I (Newton's Regime) have been plotted in Figure 12.

## Similarity with Zenz-Type Phase Diagram

From the experimental observations, it can be concluded that the phenomenon of choking or saltation is extremely gradual. A sharp transition or sudden change in pressure drop was not observed for the steady-state region. However, the acceleration zone showed behavior very similar to that suggested by Zenz [13] who suggested a sharp increase in pressure drop at low gas velocities. Such a behavior is characteristic to shorter systems. Figure 13 shows the comparison of the steady-state plot with the acceleration region plot.

## CONCLUSIONS

Experimental measurements of the static pressure profile in the direction of flow and the standard deviation in the static pressure at the entry section were made in 0.0504 m diameter set-ups held at various inclinations. The acceleration length observed in the horizontal system was significantly greater than the acceleration length observed in the vertical system. The acceleration lengths for 30 degrees set-up were almost the same magnitude as those observed in the horizontal orientation. The effect of gas velocity on acceleration length was found to be significant. At higher solids flow rates, shorter acceleration lengths were observed. The flow patterns recorded during the experiments have a strong and consistent relationship with the nature of pressure fluctuations. The variation of pressure fluctuations with gas velocity indicates that the magnitude of the fluctuations is minimum when the system is operating at the minimum pressure drop condition. From the experimental observations it was found that saltation or choking occurred first in the acceleration region. The deposition of solids slowly builds up downstream as the time progresses. Analytical expressions were derived for the Stokes' and Newton's regimes. No expression could be derived for the intermediate regime because similar mathematical simplification could not be applied.

## NOTATIONS

| | | |
|---|---|---|
| $A$ | Tube Cross-sectional Area | $(m^2)$ |
| $C_{DS}$ | Single Particle Drag Coefficient | $(-)$ |
| $C_{DM}$ | Modified Drag Coefficient, $C_D \cdot \varepsilon^{-4.7}$ | $(-)$ |
| $D_p$ | Particle Diameter | $(m)$ |
| $D$ | Tube Diameter | $(m)$ |
| $f_g$ | Gas Friction Factor | $(-)$ |
| $f_s$ | Solids Friction Factor | $(-)$ |
| $dF_d$ | Drag Force | $(N)$ |
| $dF_g$ | Gravity Force | $(N)$ |
| $dF_f$ | Friction Force | $(N)$ |
| $g$ | Gravitational Acceleration | $(m/sec^2)$ |

$K$ Parameter Group,
$$D_p \left[\frac{g \rho_f (\rho_p - \rho_f)}{\mu^2}\right]^{1/3}$$

$K_1$ Parameter Group

$$\frac{3 C_{DS} \rho_f}{4 D_p (\rho_p - \rho_f)} \quad \text{(Model I)}$$

$$\frac{18 \cdot \mu}{D_p^2 (\rho_p - \rho_f)} \quad \text{(Model II)}$$

$K_2$ Parameter Group,
$$P_k \sqrt{gD}$$

$K_3$ Parameter Group, $2K_2/D$

$K_4$ Parameter Group
$$\left\{\frac{W_s}{A \rho_p}\right\}$$

| | | |
|---|---|---|
| $L$ | Finite Length of Pipe | $(m)$ |
| $L_a$ | Acceleration Length | $(m)$ |
| $\Delta m_s, dM_s$ | Mass of the Particles in the Differential Section | $(kg)$ |

| | | |
|---|---|---|
| P | Static Pressure | (Pa) |
| $\Delta P$ | Total Pressure Drop Over Length L | (Pa) |
| $P_K$ | Constant in Konno and Saito's Equation | (-) |
| Re | Reynold Number | (-) |
| $Re_m$ | Apparent Reynold Number | (-) |

$$\varepsilon \rho_f U_g D/\mu + (1-\varepsilon)\rho_p U_p D/\mu$$

| | | |
|---|---|---|
| $Re_p$ | Particle Reynolds Number | (-) |
| $Re_t$ | Terminal Reynolds Number | (-) |
| t | Time | (sec) |
| $U_f$ | Actual Fluid Velocity, $U_g/\varepsilon$ | (m/sec) |
| $U_g$ | Superficial Gas Velocity | (m/sec) |
| $U_p$ | Particle Velocity | (m/sec) |
| $U_{p,1,2,3}$ | Roots of Cubic in eqn (2) | |
| $U_{p,a,b}$ | Roots of Quadratic in eqn (4) | |
| $U_{ss}$ | Calculated Steady State Particle Velocity | (m) |
| $U_t$ | Terminal Velocity | (m/sec) |
| $W_g$ | Gas Flow Rate | (kg/sec) |
| $W_s$ | Solid Flow Rate | (kg/sec) |

Greek Letters

| | |
|---|---|
| $\alpha$ | $K_1$ |
| $\alpha'$ | $-K_1-K_3$ |
| $\beta$ | $6.7K_1K_4-K_1(2U_g+2K_4)-K_3$ |
| $\beta'$ | $K_1K_4+K_1U_g+5.7K_1K_4-g\sin(\theta)$ |
| $\gamma$ | $K_1(U_g^2+K_4^2+2K_4U_g)-6.7K_1K_4(2U_g+2K_4)-g\sin(\theta)$ |
| $\gamma'$ | $5.7K_1K_4U_g+5.7K_1K_4^2$ |
| $\delta$ | $6.7K_1K_4(K_4^2+U_g^2+2K_4U_g)$ |
| $\varepsilon$ | Gas Voidage (-) |
| $\rho_f$ | Fluid Density (kg/m$^3$) |
| $\rho_p$ | Particle Density (kg/m$^3$) |
| $\mu$ | Fluid Viscosity (kg/m-sec) |
| $\theta$ | Angle of Inclination From Horizontal (radians) |

**LITERATURE CITED**

1. Yang, W.C. and Keairns, D., "Estimating the Acceleration Length in Vertical and Horizontal Pneumatic Transport Lines," Proc. Pneumotransport 3, Org. by BHRA Fluid Engrg., Bath, England (1976).

2. Enick, R., Falkenberg, K.L. and Klinzing, G.E., "Acceleration Length Model for Pneumatic Transport," Research Report, Univ. of Pittsburgh (1984).

3. Shi, H. and Michaelides, E.E., "The Motion of Particles in Pipelines: Collison Velocities and Acceleration Lengths," presented at the International Symposium on Multiphase Flow, Hangzhou, China (Aug. 1987).

4. Papai, L., "Examination of the Starting Section in Pneumatic Grain Conveying," Acta Technica Hungarica, 14, 95 (1956).

5. Rose, H.E. and Duckworth, R.A., "Transport of Solid Particles in Liquids and Gases," The Engineer, 28, 478 (1969).

6. Shimizu, A., Echigo, R., Hasegawa, S. and Hishida, M., "Experimental Study on the Pressure Drop and Entry Length of Gas-Solid Suspension Flow in a Circular Tube," International Journal of Multiphase Flow, 4, 53 (1978).

7. Yang, W.C., "A Unified Thoery on Dilute Phase Pneumatic Transport," Journal of Powder and Bulk Solids Technology, 1, 89 (1977).

8. Kmiec, A. and Leschonski, K., "Analysis of Models for Pneumatic Conveying of Solids in Vertical Pipes," Presented at CHISA-89, Prague (1987).

9. Michaelides, E.E., "Motion of Particles in Gases: Average Velocity and Pressure Loss," Journal of Fluids Engineering, 109, 172 (1987).

10. Konno, H. and Saito, S., "Pneumatic Conveying of Solids Through Straight Pipes," Chemical Engineering of Japan, 2(2), 211 (1969).

11. Zaltash, A., "Application of Thermodynamic Approach to Pneumatic Transport at Pipe Orientations Above the Horizontal," Ph.D. Thesis, Univ. of Pittsburgh, Pittsburgh (1987).

12. Myler, C.A., "Use of a Thermodynamic Analogy for Pneumatic Transport in Horizontal Pipes," Ph.D. Thesis, Univ. of Pittsburgh, Pittsburgh (1987).

13. Zenz, F.A., "Two Phase Fluid-Solid Flow," Ind. and Eng. Chem., 41, 2801 (1949).

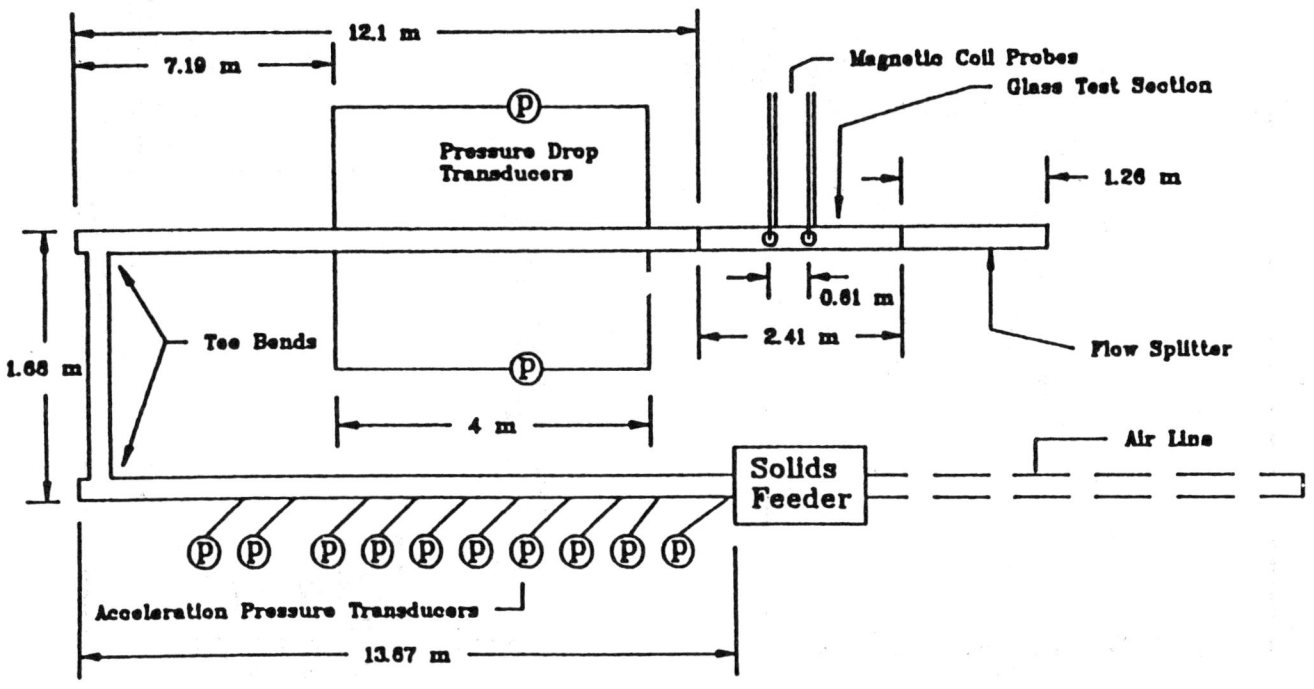

Figure 1. Layout of horizontal experimental set up.

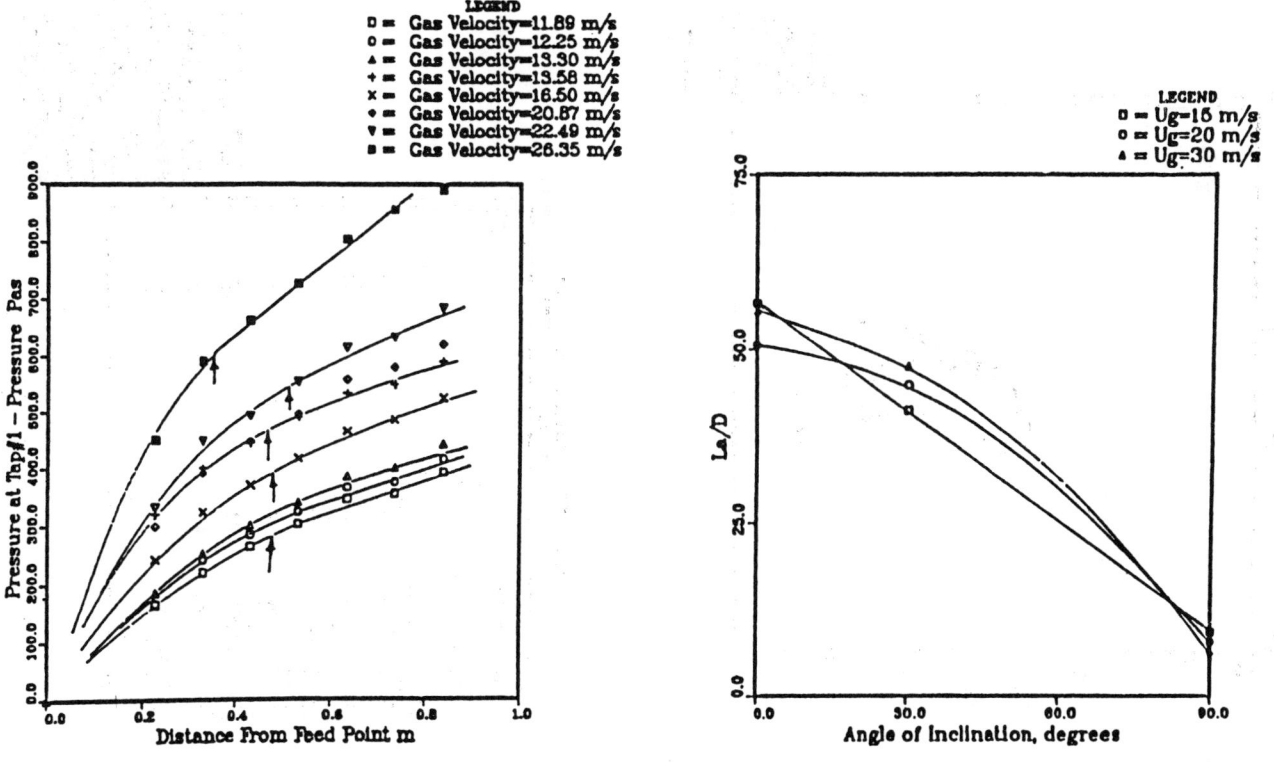

Figure 2. Representative pressure profile in the acceleration region.

Figure 3. Comparison of $L_a/D$ (average) for various inclinations at different gas velocities.

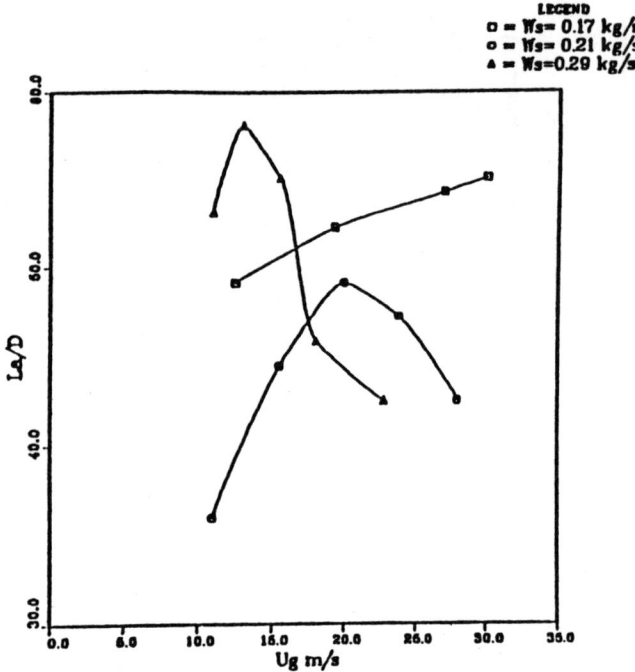

Figure 4. $L_a$ vs $U_g$ for horizontal 0.0504 m system with 450 $\mu$m glass.

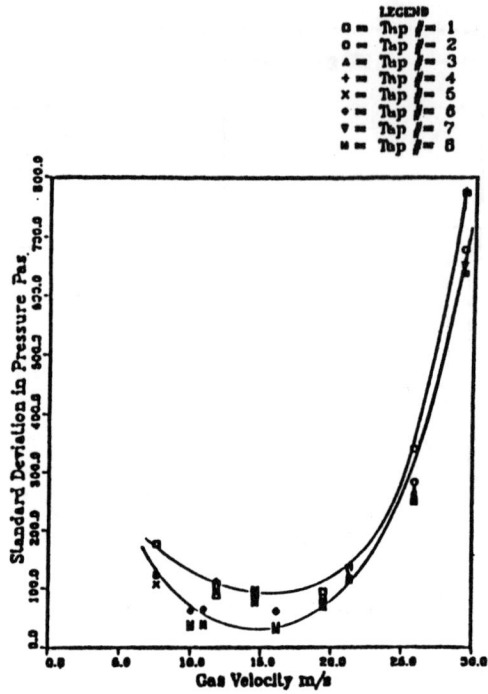

Figure 6. Standard deviation in static pressure vs gas velocity in horizontal 0.0504 m system with 450 $\mu$m glass.

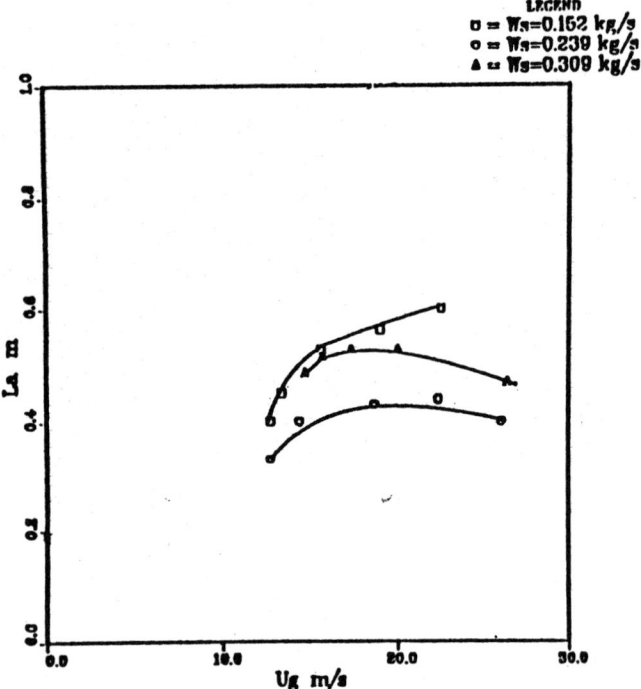

Figure 5. $L_a$ vs $U_g$ for vertical 0.0504 m system with 450 $\mu$m glass.

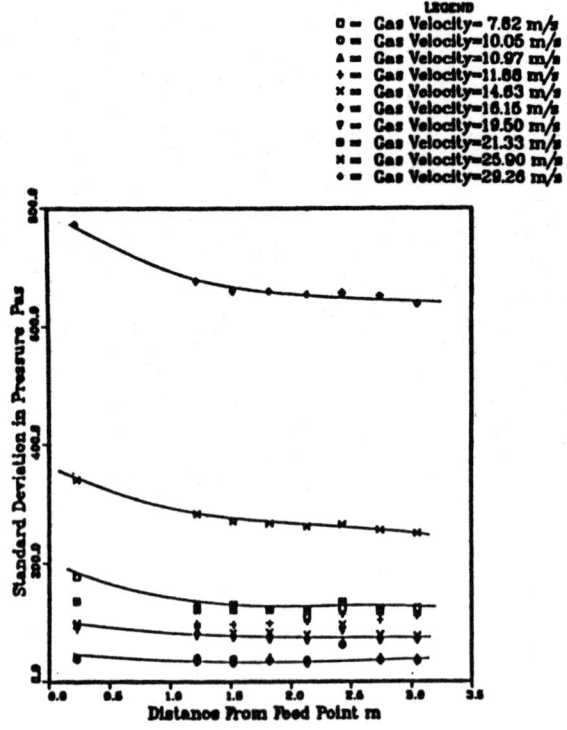

Figure 7. Variation of standard deviation in longitudinal direction for 0.0504 m horizontal system with 450 $\mu$m glass ($W_s$ = 0.11 kg/s).

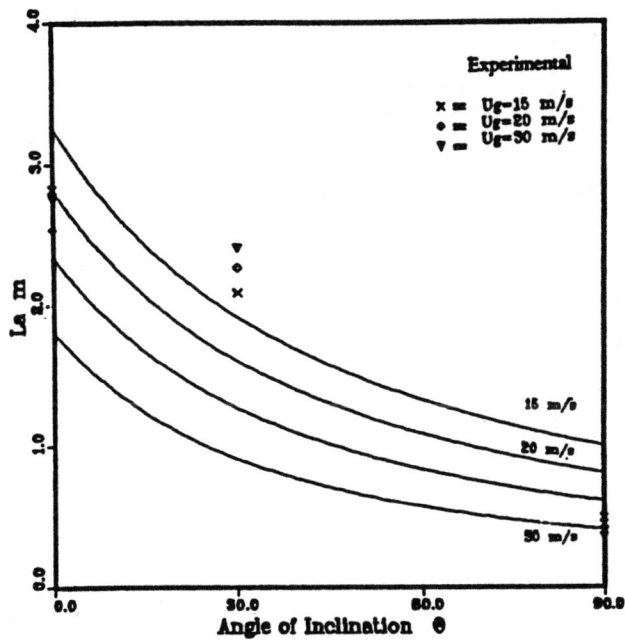

Figure 8. Comparison of experimental data (average values) with the predictions of theoretical model.

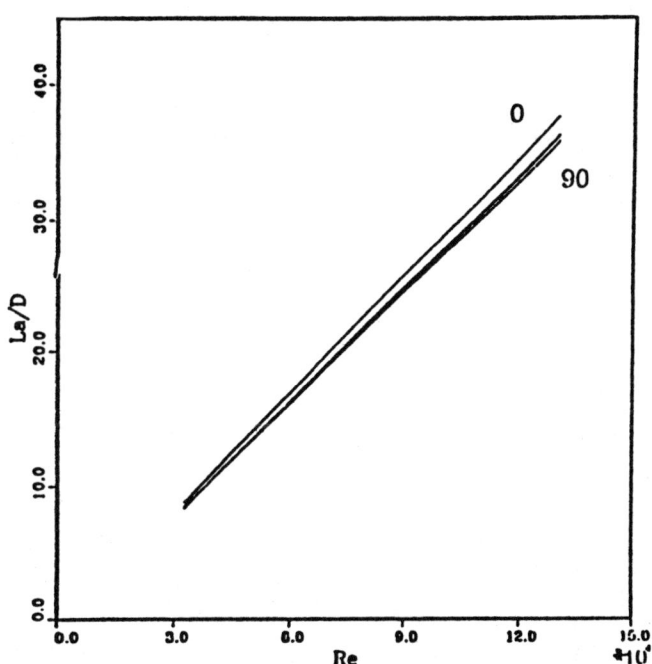

Figure 10. $L_a/D$ vs Re, as predicted by Model II for various inclinations.

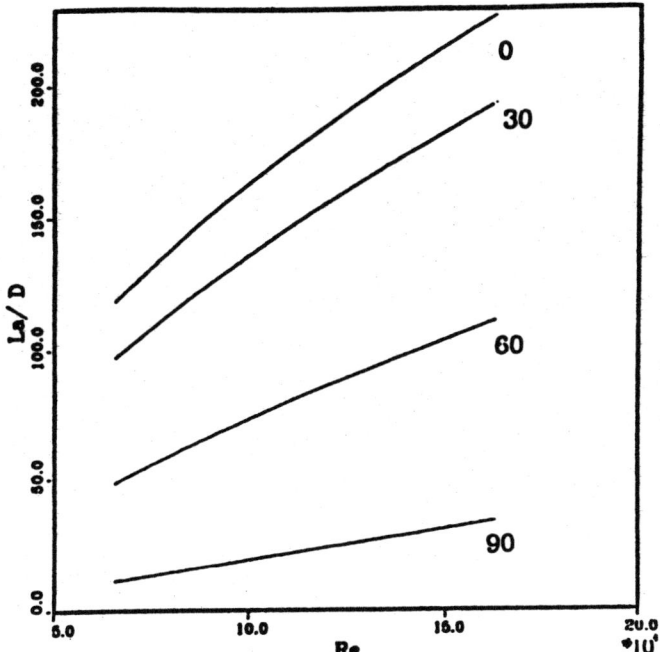

Figure 9. $L_a/D$ vs Re, as predicted by Model I for various inclinations.

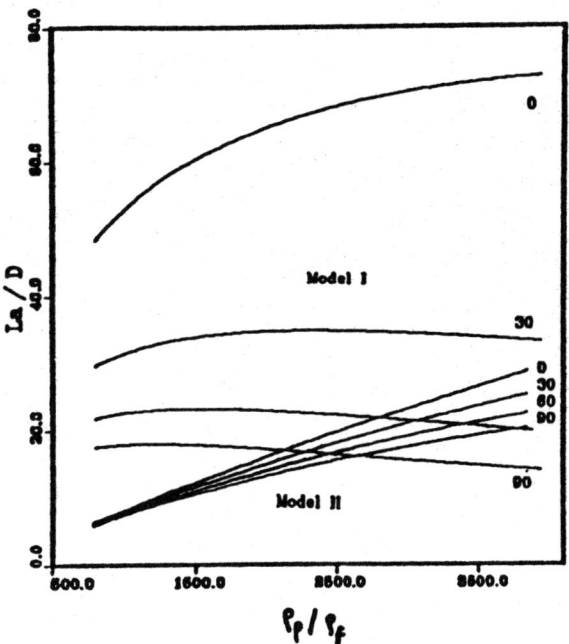

Figure 11. $L_a/D$ vs $\rho_p/\rho_f$, as predicted by Model I and II. $W_s = 0.1$ kg/s, $U_g = 15$ m/s, $D = 0.0504$ m $D_p = 450$ $\mu$m for Model I and $D_p = 50$ $\mu$m for Model II.

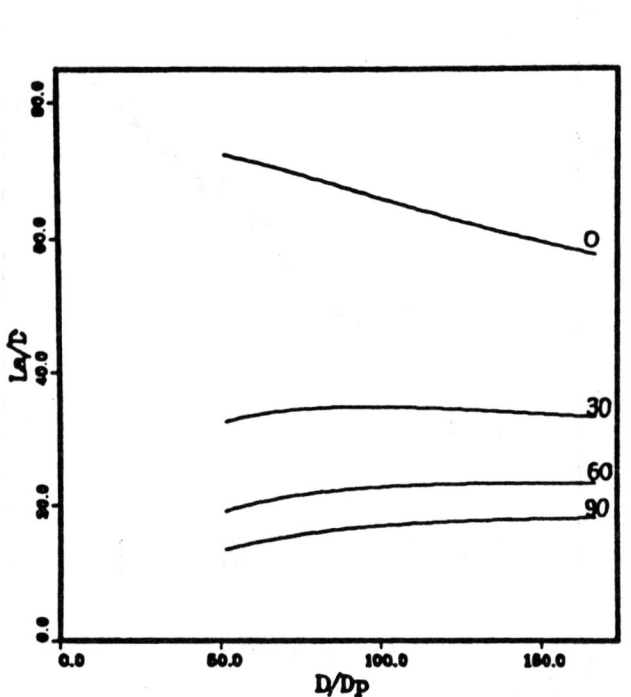

Figure 12. $L_a/D$ vs $D/D_p$, as predicted by Model I for $W_s = 0.1$ kg/s, $U_g = 15$ m/s, $\rho_p = 2400$, $\rho_f = 1.3$.

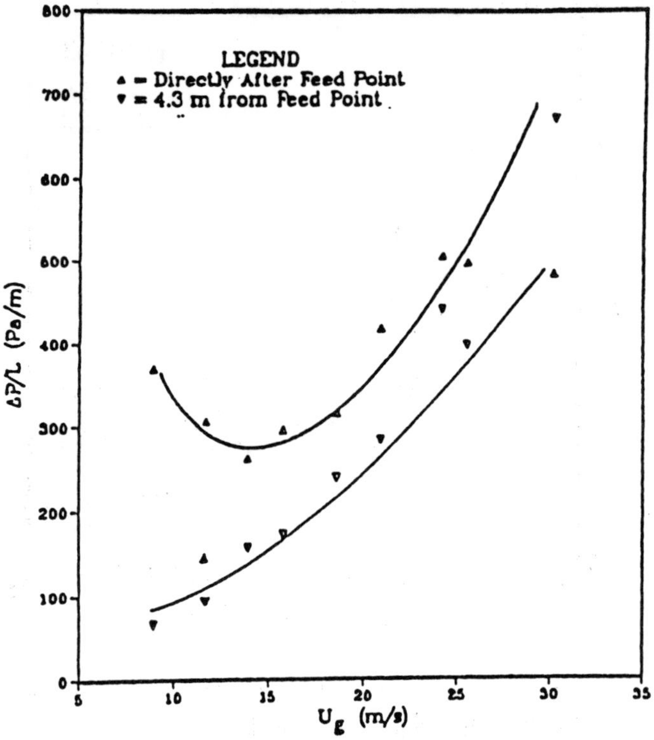

Figure 13. Comparison of $\Delta P/L$ vs $U_g$ for acceleration region and fully developed flow.

# A THEORETICAL MODEL FOR THE HYDRODYNAMICS OF GAS-SOLID TRICKLE FLOW OVER REGULARLY STACKED PACKINGS

J.H.A. Kiel and W.P.M. van Swaaij ■ Department of Chemical Engineering, University of Twente, P.O. Box 217, 7500 AE Enschede, The Netherlands

A one-dimensional steady-state model has been developed for counter-current flow of gas and solids over regularly stacked packings at dilute phase or trickle flow conditions. Starting from the separate continuity and momentum equations for both phases, expressions are derived for the average solids hold-up and the (modified) pressure drop caused by the solids. The influences of the axial packing porosity profile, the radial inhomogeneity of both the gas and solids flow and particle-particle interaction are taken into account by using two empirical parameters: 1) the solids velocity just after collision of a particle with a packing element and 2) the effective gas velocity. Proper values of these parameters can be determined independently from a comparison between calculated and experimental values of the average solids hold-up. Upon introduction of these proper values into the model, results from cold-flow experiments, in which 490 $\mu$m glass beads and air are used for the solids and gas phase respectively, can be described quite well over a wide range of solids and gas fluxes.

## INTRODUCTION

Counter-current flow of gas and solids over packed columns at dilute phase or trickle flow conditions is gaining gradually more interest as a new gas-solid contacting system. Previous studies have shown favorable properties such as a low pressure drop, low axial mixing in both phases and high rates of mass and heat transfer (1,2). Until now, industrial applications have been aimed at heat recovery from gases and particulate solids (3). However, several studies have shown the potential of gas-solid trickle flow for adsorption and chemical reactions as well (4,5).

Presently, we are investigating the application of a gas-solid trickle flow reactor as the absorber in a continuous dry regenerative process for the removal of $SO_x$ and $NO_x$ from flue gases. Copper oxide on a porous silica carrier (spherical particles) is used as the sorbent material. In flue gas treating a high gas velocity should be combined with a low pressure drop. Therefore relatively large sorbent particles are used. The hydrodynamics of gas-solid trickle flow are studied to enable prediction of important operating parameters (solids hold-up and pressure drop) and optimization of the packing configuration.

University of Twente, Enschede, The Netherlands

Experimental evidence for the possibility of stable operation of a packed column at gas-solid trickle flow conditions was given by Claus et al. (6) in a preliminary study on counter-current flow of gas and sand particles over a packing of randomly dumped cylindrical screens. They characterized gas-solid trickle flow by a relatively low solids hold-up and pressure drop. The presence of the packing reduces the pressure drop caused by the solids to a large extent, because the solids are actually supported by the packing. Three different regimes of operation are distinguished, just like in gas-liquid trickle flow. At low gas velocities, a constant particle velocity is found (preloading), while at higher gas velocities the particle velocity decreases causing an increase in solids hold-up. In this way loading occurs. Eventually, if the gas velocity is further increased, transition to a more dense phase regime is reached (flooding).

The hydrodynamics of fine powders (Fluid Cracking Catalyst (FCC)) flowing over randomly dumped packings have been studied in more detail by Roes and van Swaaij (7) and by Westerterp and Kuczynski (8). In both studies the modelling is based mainly on similarities between gas-solid and gas-liquid trickle flow. Next to FCC, Verver and van Swaaij (9) also used coarse particles like sand and steel shot. They applied a regularly stacked packing of bars with a square cross-section, in stead of a dumped packing, to minimize pressure drop and static solids hold-up. Their hydrodynamic model has been developed for coarse particles. It is based on the momentum equation of a single particle and particle-particle interaction is assumed to have no influence on hold-up and pressure drop. For trickle flow of coarse particles (425 μm sand, and 330 and 880 μm steel shot), the experimental data on solids hold-up and pressure drop can be described quite well by this model. However, the hydrodynamic behavior for 70 μm FCC and 225 μm sand seems to be considerably influenced by particle-particle interaction.

The present paper reports on further improvement of Verver and van Swaaij's model. The steady-state one-dimensional hydrodynamic model presented hereafter is now systematically derived from the separate continuity and momentum equations for both the gas and solids phase. Additionally, the influences of the axial packing porosity profile, the radial inhomogeneity of both the gas and solids flow, and particle-particle interaction are taken into account. The modelling results are then compared with results of measurements in a cold-flow set-up with a 0.10x0.10x0.50 m³ gas-solid trickle flow column (made of transparent lexan to allow visual observations). The column packing consisted of cross-wise stacked PVC bars with a circular cross-section (see Figure 1). In the experiments, 490 μm glass beads have been used for the solids phase and air for the gas phase.

MODEL DEVELOPMENT

Gas-solid trickle flow, being a dilute flow of solid particles, is comparable with pneumatic conveying. Differences in hydrodynamic modelling are caused mainly by the presence of a packing in case of gas-solid trickle flow. For instance, particle acceleration is not merely an entrance effect, like in pneumatic conveying, but it is important throughout the entire column.

In the following section, the steady-state one-dimensional hydrodynamic model for gas-solid trickle flow will be described. Starting from the separate continuity and momentum equations for both the gas and solids phase, expressions will be derived for two important design parameters, the average solids hold-up and the pressure drop caused by the solids phase. The modelling is focussed on the flow phenomena inside a single packing cell (see Figure 1). As a result, the expressions for both parameters are independent of the column dimensions. The influences of i) the axial gas velocity profile caused by the packing, ii) the inhomogeneity of both the radial gas and solids distribution and iii) particle-particle interaction, are included in the value of two special parameters which have to be determined experimentally. Basic assumptions of the steady-state one-dimensional hydrodynamic model are:
- The particles are spherical and uniform in size and density.
- The particles are considered to constitute a particulate phase; both, gas and particulate phase form a continuum.
- The pressure gradient exists in the gas phase only.
- The packing porosity is constant over the column length.
- The interaction between gas and packing plus column wall does not depend on the presence of solids.

The conservative and constitutive equations resulting from these assumptions are presented below. It should be noted that the axial co-ordinate is positive in downward direction. Consequently, for counter-current operation, the upwards directed gas velocity should be taken negative.

Continuity equations

gas phase:
$$\frac{dG}{dz} = \frac{d}{dz}\left[(\varepsilon-\beta)\rho_g u_g\right] = 0 \quad (1)$$

solids phase:
$$\frac{dS}{dz} = \frac{d}{dz}(\beta\rho_s u_s) = 0 \quad (2)$$

Momentum equations

gas phase:
$$(\varepsilon-\beta)\rho_g u_g \frac{du_g}{dz} = -\varepsilon\frac{dP}{dz} + F_{Gg} + F_{Fg} + F_D \quad (3)$$

solids phase:
$$\beta\rho_s u_s \frac{du_s}{dz} = F_{Gs} + F_{Fs} - F_D \quad (4)$$

The buoyancy force has been neglected, because its value scales, in comparison to the other terms in Equation (4), as the gas-solids density ratio.

In dilute gas-solids flow, the gas density $\rho_g$ is usually considered to be constant, because the total pressure drop is small compared to the absolute pressure level. Also, the packing porosity $\varepsilon$ is constant over the column length and always much larger than the solids hold-up $\beta$. Therefore (see Equation (1)), the gas velocity must be practically constant. The gas phase continuity equation can then be reduced to:

$$u_g = \text{constant} = \frac{G}{(\varepsilon - \bar{\beta})\rho_g} \approx \frac{G}{\varepsilon \rho_g} \quad (5)$$

where $\bar{\beta}$ and $\rho_g$ represent averaged values for the solids hold-up and the gas density over the height of the column. If $u_g$ is taken to be constant, the convective acceleration term in Equation (3) should accordingly be zero, by which the momentum equation for the gas phase is simplified to:

$$\varepsilon \frac{dP}{dz} = F_{Gg} + F_{Fg} + F_D \quad (6)$$

For a further treatment of the momentum equation for the solids phase, the concept of a packing cell is introduced. Such a packing cell (see Figure 1) is the smallest basic unit of the column, with the dimensions of $L_v$ (height) and $0.5 L_h$ (width). A regularly stacked column can then be considered as a two-dimensional array of packing cells, and the hydrodynamic modelling be restricted to a single cell. Consequently, the model becomes independent of the column dimensions.

The "single packing cell" approach is allowed if radial segregation of gas and solids on the scale of the column diameter, as well as the interaction of the solids with the column wall, are unimportant. It is further assumed that the particles collide only once with the packing element of a cell and that the solids velocity profile is the same in each horizontal packing layer.

The above assumptions with the single packing cell concept are confirmed by visual observations during cold flow experiments in a 0.10x0.10x0.50 m³ column made of transparent lexan. Different regularly stacked packings were used and glass beads were taken as the solids phase material. It appears, that radial segregation is only a minor effect and that the particles do collide with a packing element of each packing layer indeed. Experimental evidence for the solids velocity profile being the same in each packing layer is given by Guigon (10). He measured the total pressure drop as a function of the column height for a packing of several layers of horizontal tubes in a staggered configuration and found a linear relationship. Evidently, the additional assumptions do not seem to restrict the applicability of the model severely.

The solids phase momentum equation applied for the period between two successive collisions with a packing element can now be written as:

$$\beta \rho_s u_s \frac{du_s}{dz} = F_{Gs} - F_D \quad (7)$$

The solids-packing interaction is then implied as the initial solids velocity $u_{so}$, in stead of being described as a friction force acting on the solids during their way through the column. The initial solids velocity is defined as the solids velocity immediately after collision with a packing element. For a single particle, it would basically be possible to express $u_{so}$ as a function of the solids velocity just before collision with a packing element ($u_{sLv}$). However, the collision is generally influenced by the presence of other particles, wich makes the relation between $u_{sLv}$ and $u_{so}$ very complex. Therefore, in the model, the initial solids velocity will be treated as an empirical input parameter.

Constitutive equations

To solve Equations (2), (6) and (7), the constitutive equations for the various forces must be known. The gravity forces $F_{Gg}$ and $F_{Gs}$ are given by:

$$F_{Gg} = (\varepsilon - \beta)\rho_g g \quad (8)$$
$$F_{Gs} = \beta \rho_s g \quad (9)$$

and the friction force $F_{Fg}$ exerted by the packing and the column wall on the gas phase can, at turbulent flow conditions, be expressed as:

$$F_{Fg} = - K \frac{1}{2} \rho_g u_g |u_g| \quad (10)$$

Because the concentration of solids in gas-solid trickle flow is quite small, it may be assumed in first approximation, that the flow field around each individual particle is not influenced by the presence of surrouding particles. Then, the solids flow may be considered as single particle flow and the drag force can be written as:

$$F_D = \beta C_{Ds} \frac{3}{4} \frac{\rho_g}{d_s} (u_s - u_g)|u_s - u_g| \quad (11)$$

where $C_{Ds}$ is the drag coefficient of a single sphere. According to the widely used correlation of Schiller and Neumann (11), $C_{Ds}$ can be calculated from:

$$C_{Ds} = \frac{24}{Re}(1 + 0.15 Re^{0.687}) \text{ for } Re < 10^3 \quad (12)$$

where:

$$Re = \frac{\rho_g d_s |u_s - u_g|}{\mu_g} \quad (13)$$

In contrast with the assumption discussed above, the influence of surrounding particles on the flow field around a certain particle can normally not be neglected. It will be described later, how this influence is included in the model.

## Average solids hold-up

The hydrodynamic model presented above allows the calculation of the local solids hold-up $\beta$. However, for design purposes it is usually sufficient to know the average solids hold-up $\bar{\beta}$. An expression for $\bar{\beta}$ can be derived by integrating the continuity equation for the solids phase over the height $L_v$ of a packing cell. Two situations should be distinguished, because the initial solids velocity may either be positive or negative. A negative initial solids velocity means, that the solids move in upward direction immediately after collision with a packing element.

Figure 1 shows that, when the initial solids velocity is negative, the solids hold-up just above a packing element consists of three different contributions. The average solids hold-up in a packing cell may then be written as:

$$\bar{\beta} = \frac{1}{L_v} \int_0^{L_v} \beta \, dz$$

$$= \frac{S}{\rho_s L_v} \int_0^{L_v} \frac{1}{|u_{s1}|} dz + \int_{L_*}^{L_v} \left( \frac{1}{|u_{s2}|} + \frac{1}{|u_{s3}|} \right) dz$$

$$= \frac{St_{L_v}}{\rho_s L_v} = \frac{S}{\rho_s \bar{u}_s} \qquad (14)$$

If the solids velocity profile is assumed to be the same in each packing cell, $\bar{\beta}$ also represents the average solids hold-up of the entire packed section. The second integral in Equation (14) is zero in case of a positive initial solids velocity. However, the relation between $\bar{\beta}$ and $\bar{u}_s$ remains the same.

## (Relative) pressure drop

Another important hydrodynamic design parameter for gas-solid trickle flow reactors is the pressure drop. If the interaction between gas and packing is indeed independent of the presence of solids, the pressure drop over the height of a packing cell can be derived from the integrated momentum equation (6) for the gas phase as:

$$\Delta P = \Delta P_p + \Delta P_s \qquad (15)$$

The total pressure drop equals the sum of the pressure drop caused by friction between gas and packing plus column wall ($\Delta P_p$) on the one hand, and friction between gas and solids ($\Delta P_s$) on the other. The pressure drop due to the gravity force has not been taken into account because, upon experimental verification, this pressure drop will usually not be measured.

$\Delta P_p$ can easily be calculated from Equation (10) for turbulent flow conditions, which prevail in usual applications of gas-solid trickle flow. The value of the constant K depends on the packing configuration. For certain configurations, existing correlations can be used to estimate the value of K (12). Otherwise the value of K can easily be determined experimentally. The pressure drop due to friction between gas and solids ($\Delta P_s$) is more complex. Next to the average solids hold-up, it is the second important output parameter of the hydrodynamic model.

$\Delta P_s$ is generally influenced by the presence of a packing. A dimensionless parameter will now be introduced to quantify this influence. If no packing is present, like in pneumatic conveying, particle acceleration usually exists only as an entrance effect. In the major part of the column the solids flow with a steady-state velocity and the drag force equals the gravity force acting on the solids (see Equation (7)). In that case the pressure drop due to friction between gas and solids reaches its maximum value for the adjusted gas and solids mass fluxes. In gas-solid trickle flow the packing reduces the velocity of the solids and therefore the drag force (Equation (11)). Consequently, the pressure drop due to friction between gas and solids is also reduced. According to Verver and van Swaaij (9), this reduction can be quantified by the introduction of a relative pressure drop $\gamma$, which is defined as the quotient of the drag force, exerted by the solids on the gas, and the gravity force on the solids:

$$\gamma = \frac{F_D}{\beta \rho_s g} \qquad (16)$$

Experimental determination is possible if the average relative pressure drop over a single packing cell, which is assumed to be equal to the average relative pressure drop over the entire packing, is defined as:

$$\bar{\gamma} = \frac{\bar{F}_D}{\bar{\beta} \rho_s g} \qquad (17)$$

Because the average drag force $\bar{F}_D$ in the last equation can be substituted by $\Delta P_s \varepsilon / L_v$ (combine Equation (15) with the integrated form of Equation (6)), the average relative pressure drop can also be written as:

$$\bar{\gamma} = \frac{\Delta P_s \epsilon}{\bar{\beta} \rho_s g L_v} \qquad (18)$$

Herein $\epsilon$, $L_v$ and $\rho_s$ are known properties of the packing and the solids respectively, while $\Delta P_s$ and $\bar{\beta}$ can be determined experimentally.

A theoretical expression for the average drag force can be derived by combining the momentum equation for the solids phase (7) with Equation (9) for the gravity force and the integrated form of Equation (2) (that is: $\beta \rho_s u_s = S$), according to:

$$\bar{F}_D = \frac{1}{L_v} \int_0^{L_v} (\rho_s g - \rho_s u_s \frac{du_s}{dz}) \frac{S}{\rho_s u_s} dz$$

$$= \frac{S}{L_v} [g t_{L_v} - (u_{sLv} - u_{so})] \qquad (19)$$

$u_{sLv}$ represents the solids velocity just before, and $u_{so}$ the solids velocity just after collision with a packing element. Substitution of Equations (14) and (19) into (17) finally gives:

$$\bar{\gamma} = 1 - \frac{1}{g t_{L_v}} (u_{sLv} - u_{so}) \qquad (20)$$

If particle acceleration is absent, then $\bar{\gamma} = 1$ ($u_{sLv} = u_{so}$); otherwise $\bar{\gamma} < 1 (u_{sLv} > u_{so})$. In Equation (20) $\bar{\gamma}$ is expressed as a function of the solids velocity profile, just like $\bar{\beta}$ in Equation (14). As described hereafter, the solids velocity profile is found by solving Equation (7). First however, several phenomena will be discussed, which affect the drag force exerted by the solids on the gas, and therewith the relative pressure drop $\bar{\gamma}$.

Effective gas velocity

The drag force $F_D$, exerted by the solids on the gas at a certain axial position z (Equation (11)), represents the sum of the drag forces exerted by all the individual particles at that same axial position. Because the model is one-dimensional, radially averaged values of the local hold-up and the local gas and solids velocity are required for Equation (11). In first approximation, the local solids velocity $u_g$ was supposed to along the column and equal to $G/(\epsilon \rho_g)$. There may be three reasons for the (radially averaged) local gas velocity to deviate from this value.

First, the actual packing porosity is not constant along the column. For a regularly stacked packing it is easily possible to eliminate this problem by expressing the packing porosity as a function of the axial co-ordinate and correct the local gas velocity for this actual packing porosity. However, generally this correction for the local gas velocity is not sufficient to yield the actual drag force, because a second effect, namely the inhomogeneity of both the radial gas and solids distribution in the free space of the column, plays an important role. The inhomogeneity of the gas flow is caused by viscous forces. Boundary layers with a velocity gradient are developed around each packing element, and for turbulent flow (prevailing in usual applications of gas-solid trickle flow) large wakes exist beyond the packing elements. If the gas flow is inhomogeneous, the solids flow will also be inhomogeneous. As stated before, the drag force should actually be based on the relative velocities of individual particles. But due to the complexity of the above described phenomena it is hardly possible to find a precise theoretical expression for the radially averaged relative velocity in Equation (11). Therefore, in the present model an effective gas velocity is used, given by the following equation:

$$u_{geff} = \xi \frac{G}{\epsilon \rho_g} \qquad (21)$$

$\xi$ is an empirical parameter, which is indicated as the "effectiveness factor".

Except for the factors mentioned, the drag force should also be corrected for the influence of particle shielding. Verver and van Swaaij (9) clearly showed this influence to occur for gas-solid trickle flow of 70 μm Fluid Cracking Catalyst (FCC) as well as 255 and 425 μm sand particles. They found the drag force to decrease with increasing solids hold-up. Therefore, a Richardson and Zaki type correlation (13) is not appropriate, because it predicts the opposite effect of an increasing drag force with increasing solids hold-up. For FCC, Verver and van Swaaij showed that a correlation proposed by Matsen (14) can be successfully applied if the influence of particle acceleration is small. However, a general correlation for particle shielding in gas-solid trickle flow applicable for different particle diameters and densities, does not exist. Besides, it is difficult to distinguish experimentally between the shielding effect and the effects of the axial packing porosity profile and radial inhomogeneities in gas and solids flow. Therefore, in the present model, the effect of particle shielding is additionally included in the previously defined effectiveness factor $\xi$.

## Numerical solution

Calculation of the average solids hold-up $\bar{\beta}$ and the average relative pressure drop $\bar{\gamma}$ from Equations (14) and (20) requires the following parameters to be known: i) the time averaged solids velocity $\bar{u}_s$, ii) the solids velocity just before and just after collision with a packing element ($u_{sLv}$ and $u_{so}$) and iii) the time between two successive collisions $t_{Lv}$. These parameters in turn can be evaluated from the solution of the momentum equation for the solids phase (7) together with the constitutive equations (9) and (11). If, in Equation (7), $u_s/dz$ is substituted by dt, the problem is reduced to a straightforward initial value problem, with $u_s = u_{so}$ at $t = 0$:

$$\beta \rho_s \frac{du_s}{dt} = F_{Gs} - F_D \qquad (22)$$

$F_D$ is a function of the effective gas velocity, i.e. the empirical parameter $\xi$. Both $u_{so}$ and $\xi$ should have a certain value a priori, which can be found by fitting the results from the model to experimental data.

Equation (22) has to be solved numerically, for instance by a 4th order Runge-Kutta forward integration method. To determine the time at which a particle reaches the next packing element, the following equation has to be solved simultaneously. It relates the axial position of a particle to the falling time after the last collision:

$$\frac{dz}{dt} = u_s \qquad (23)$$

The initial condition for Equation (23) is $z = 0$ at $t = 0$. Since this system of two first order ordinary differential equations is stiff, the time step in the Runge-Kutta method should be very small. Therefore, a Fortran library routine (NAG-library) is used, which is especially developed for such systems. In this way, the number of calculation cycles could be reduced by a factor 20-40 if compared to the straightforward Runge-Kutta method.

## MODEL COMPUTATIONS

For model computations, values of the following input parameters should be known: i) gas and solids mass flux, ii) density and viscosity of the gas, iii) average density and diameter of the solid particles and iv) porosity and vertical pitch of the packing. Values of the two empirical parameters, $u_{so}$ and $\xi$, should also be set a priori. In fact, $u_{so}$ and $\xi$ are a function of both the gas and solids mass fluxes for a certain gas-solids-packing combination. However, in the model computations their values are assumed to depend on the solids mass flux only.

A trial and error method has to be applied to find the values of both empirical parameters, for which calculated and experimental values of the two output parameters are in best agreement. From Figure 2, it appears that values for $u_{so}$ and $\xi$ can be determined independently using only the hold-up versus gas mass flux plot. Near G = 0, the drag force is small compared to the gravity force. Because $\xi$ appears in the drag force only, the axial solids velocity profile and consequently the calculated average solids hold-up are then hardly influenced by it. Therefore the proper value of $u_{so}$ can be determined at low gas mass fluxes by fitting the first part (G → 0) of a calculated $\bar{\beta}$ vs. G curve to the experimentally obtained one. In the loading regime (i.e. for high gas mass fluxes), the influence of $\xi$ is dominating, the drag force being of the same order of magnitude as the gravity force. So, the proper value of $\xi$ can be determined by matching the last part of calculated and experimental $\bar{\beta}$ vs. G curves. Since $\bar{\gamma}$ is not used to determine the proper values of both empirical parameters, comparison of calculated and experimental values of $\bar{\gamma}$ directly reflects the quality of the model.

Figure 3 gives a comparison of experimental results and model computations. The experimental results are obtained from cold flow experiments in a 0.10x0.10x0.50 $m^3$ column made of lexan with a packing of cross-wise stacked bars with a circular cross-section (see Figure 1). The bars are made of PVC and 10 mm in diameter. For the solids phase, 490 μm glass beads are used, and air for the gas phase. The experimental conditions are listed in Table 1. In the model computations, $u_{so}$ is directed upwards and kept constant at -0.05 m/s. Three different modifications of computed results are presented in Figure 3, according to curves a, b and c. Each curve refers to another way of modelling the gas-solids interaction.

For curve a, no corrections have been made at all for any influences of the axial packing porosity profile, inhomogeneities in the radial gas and solids distribution, or particle shielding ($u_g$ = constant, $\xi = 1.0$). It appears that in this approach calculated values of both output parameters are too low, especially at high gas mass fluxes. This indicates that at least one of the influences mentioned cannot be neglected.

For curve b, the local gas velocity has

been corrected for the axial packing porosity profile only, according to:

$$u_g = \frac{G}{\rho_g} \left[ 1 + \left| \sin\left(\frac{2\pi}{D}\right) \right| \right] \quad (24)$$

Compared to case a, this results in smaller values of $\bar{\beta}_{calc}$ and $\bar{\gamma}_{calc}$ over a wide range of gas mass fluxes. Apparently, the agreement between calculated and experimental values is not improved by this correction. This is caused by the fact that for this regime of gas mass fluxes, the main part of the solids hold-hold-up is situated just above the packing elements. In case b, the packing porosity just above a packing element is higher than in case a, which results in a lower local gas velocity and thus in lower values of $\bar{\beta}_{calc}$ and $\bar{\gamma}_{calc}$. For very high gas mass fluxes, $\bar{\beta}$ and $\bar{\gamma}$ are mainly determined by the gas velocity in the minimum free cross-sectional area of the column. If this gas velocity is approaching the terminal velocity of the solids, it becomes difficult for them to pass this minimum free area and then, the main part of the solids hold-up will be situated just above it. Since for this axial position, the packing porosity in case b is lower than in case a, this results in larger values of $\bar{\beta}$ and $\bar{\gamma}$ in approach b at high gas mass fluxes.

Finally, for curve c the influences of all the phenomena discussed have been taken into account by allowing the value of the effectiveness factor $\xi$ to differ from one. With this approach, the agreement between calculated and experimental values is good over the entire range of gas mass fluxes in the preloading and loading regime for $\xi = 1.35$. This means that the effective gas velocity $u_{geff}$ is higher than the superficial gas velocity corrected for the average packing porosity. Because the influence of particle shielding alone would yield a decreased $u_{geff}$ ($\xi < 1$), here the radial inhomogeneity of the gas and solids flow must be the dominating phenomenon. In the loading regime, the gradient of $\bar{\beta}_{calc}$ is higher than the gradient of $\bar{\beta}_{exp}$. This is probably caused by the fact, that in the experiments the particle diameter is distributed over a certain range, while in the model computations a uniform particle size is assumed (experimentally $d_s = 420 - 590$ μm and in the model $d_s = 490$ μm).

To illustrate the accuracy of the hydrodynamic model (modification c), calculated versus experimental values of $\bar{\beta}$ and $\bar{\gamma}$ are shown in Figure 4. With the proper values of $u_{so}$ and $\xi$, the maximum error in $\bar{\beta}$ is about 10%. $\bar{\gamma}$ can usually be predicted with an error of at most 20%. However, for low values of $\bar{\gamma}$ (that is for relatively low gas mass fluxes), the calculated values are up to 40% too small.

This may be caused, in the first place, by backmixing in the gas phase. Gas recirculation occurs if part of the gas is dragged downwards by the solids flow and released again by collision of the particles with a packing element. As a consequence an additional contribution exists to both, the pressure drop caused by friction between gas and solids ($\Delta P_s$), and the pressure drop due to friction between gas and packing plus column wall ($\Delta P_p$). These contributions are not included in the model. Neither has been taken account of any influence of the presence of solids on ($\Delta P_p$) by a possible effect on the radial gas velocity distribution.

A second explanation for the deviation between $\bar{\gamma}_{calc}$ and $\bar{\gamma}_{exp}$ may be found in the decreasing influence of particle shielding with decreasing gas mass fluxes. The effectiveness factor $\xi$, which accounts for this effect, has been evaluated at high values of G (loading regime) but then applied over the entire range of G. It can be expected however, that the particle shielding effect becomes smaller for lower values of G because of the corresponding decrease in solids hold-up. As a consequence, the model underestimates the relative pressure drop at low values of G.

## CONCLUDING REMARKS

For counter-current flow of 490 μm glass beads and air over a regularly stacked packing, the presently developed hydrodynamic model describes two important design parameters, viz. the average solids hold-up $\bar{\beta}$ and the relative pressure drop caused by the solids $\bar{\gamma}$, quite well in the preloading and loading regime.

The influences of the axial packing porosity profile, the radial inhomogeneity of the gas and solids flow and particle shielding cannot be neglected. In the model, these influences are involved in two empirical parameters, the initial solids velocity just after collision of a particle with a packing element $u_{so}$ and the effectiveness factor $\xi$. Both parameters are assumed to be independent of the gas mass flux G. Proper values of $u_{so}$ and $\xi$ can be determined independently from a comparison between calculated and experimental values of the average solids hold-up.

A detailed experimental study of the hydrodynamic behavior for different packing configurations will soon be published. In that study, both empirical parameters $u_{so}$ and $\xi$ will be further considered as a function of packing configuration, solids mass flux and particle diameter.

ACKNOWLEDGEMENTS

The authors wish to thank G.G. Oosterwegel for the stimulating discussions on the modelling concept and W. Prins for his valuable comments on the manuscript.

NOTATION

Roman letters

$C_{Ds}$ = drag coefficient
$d_s$ = average particle diameter [m]
$D$ = diameter of a packing element [m]
$F_D$ = drag force per unit reactor volume [N/m³]
$F_F$ = friction force per unit reactor volume [N/m³]
$F_G$ = gravity force per unit reactor volume [N/m³]
$g$ = acceleration of gravity [m/s²]
$G$ = gas mass flux [kg/(m²s)]
$K$ = drag constant for packing/wall [1/m]
$L$ = column length [m]
$L^*$ = bouncing height [m]
$L_h$ = horizontal pitch [m]
$L_v$ = height of packing layer or vertical pitch [m]
$P$ = pressure [Pa]
$Re$ = particle Reynolds number
$S$ = solids mass flux [kg/(m²)]
$t$ = time [s]
$U_g$ = superficial gas velocity [m/s]
$u_g$ = local gas velocity [m/s]
$u_{geff}$ = effective local gas velocity [m/s]
$u_s$ = local solids velocity [m/s]
$\bar{u}_s$ = time averaged solids velocity [m/s]
$u_t$ = particle terminal velocity [m/s]
$z$ = axial co-ordinate [m]

Greek letters

$\beta$ = local solids hold-up [m³/(m³reactor)]
$\bar{\beta}$ = average solids hold-up [m³/(m³reactor)]
$\gamma$ = local relative pressure drop of solids (Equation (16))
$\bar{\gamma}$ = average relative pressure drop of solids (Equation (17))
$\varepsilon$ = void fraction of the packing
$\mu$ = dynamic viscosity [Pa s]
$\xi$ = effectiveness factor (Equation (21))
$\rho$ = density [kg/m³]

Subscripts

g = gas phase quantities
Lv = conditions at the end of a packing cell
o = conditions at the beginning of a packing cell
p = packing quantities
s = solids phase quantities

LITERATURE CITED

1. Roes, A.W.M. and W.P.M. van Swaaij, Chem. Eng. J., 18, 29 (1979).

2. Verver, A.B. and W.P.M. van Swaaij, Powder Technol., 45, 133 (1986).

3. Guigon, P., J.F. Large and Y. Molodtsof, "Hydrodynamics of raining packed bed heat exchangers", Vol. 4, Ch. 39 of "Encyclopedia of fluid mechanics", N.P. Cheremisinoff ed., Gulf Publ. Co., Houston (1986).

4. Kuczynski, M.: The synthesis of methanol in a gas-solid-solid trickle flow reactor, PhD thesis University of Twente, the Netherlands (1986).

5. Verver, A.B. and W.P.M. van Swaaij, Chem. Eng. Sci., 42 (3), 435 (1987).

6. Claus, G., F. Vergnes and P. le Goff, Can. J. Chem. Eng., 54, 143 (1976).

7. Roes, A.W.M. and W.P.M. van Swaaij, Chem. Eng. J., 17, 81 (1979).

8. Westerterp, K.R. and M. Kuczynski, Chem. Eng. Sci., 42 (7), 1539 (1987).

9. Verver, A.B. and W.P.M. van Swaaij, Powder Technol., 45, 119 (1986).

10. Guigon, P., "Hydrodynamique d'un lit ruisselant de particules en transfert thermique associé application au dimensionnement d'un échangeur industriel, PhD thesis Compiègne University of Technology, France (1987).

11. Schiller, L. and A. Neumann, Z. Ver. dtsch. Ing., 77, 318 (1933).

12. Perry, R.H. and D. Green, Perry's Chemical Engineers' Handbook, McGraw-Hill, New York (1984).

13. Richardson, J.F. and W.N. Zaki, Trans. Instn. Chem. Engrs., 32, 35 (1954).

14. Matsen, J.M., Powder Technol., 32, 21 (1982).

Table 1. Experimental conditions.

| | |
|---|---|
| solids mass flux $S$ [kg/(m²s)] | 1.0 |
| gas density $\rho_g$ [kg/m³] | 1.2 |
| dynamic viscosity of the gas $\mu_g$ [Pa s] | $1.8\ 10^{-5}$ |
| solids density $\rho_s$ [kg/m³] | 2480 |
| average diameter of the solids $d_s$ [μm] | 490 |
| diameter of a packing element $D$ [m] | 0.010 |
| packing porosity $\varepsilon$ | 0.607 |
| vertical pitch $L_v$ [m] | 0.010 |
| horizontal pitch $L_h$ [m] | 0.020 |

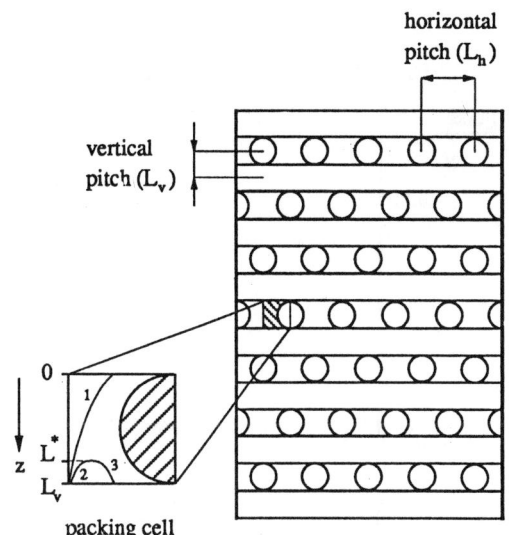

Figure 1. Packing configuration, tested in a 0.10x0.10x0.50 m³ column, and trajectory of a single particle in a packing cell.

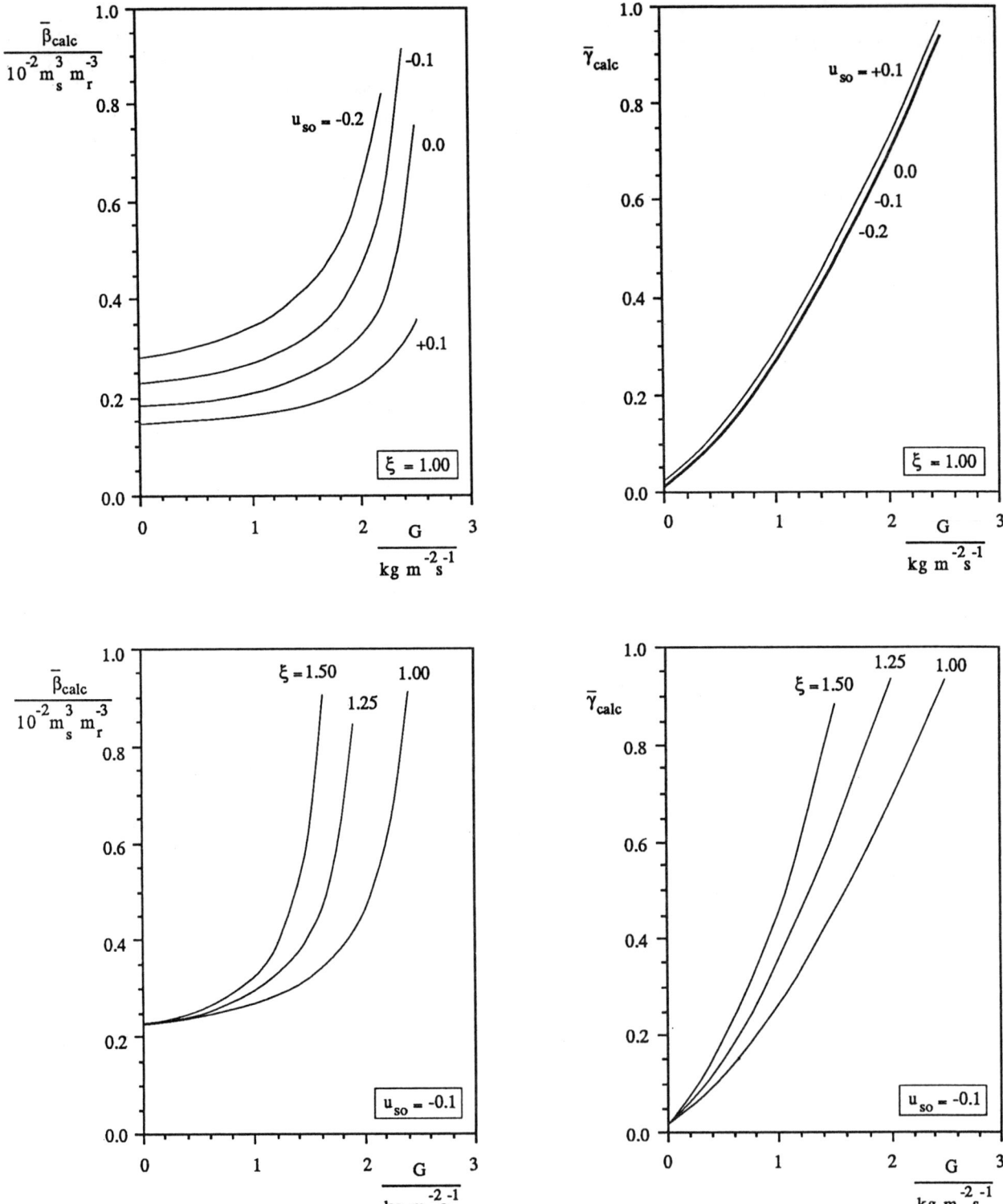

Figure 2. Influence of initial solids velocity $u_{so}$ and effectiveness factor $\xi$ on average solids hold-up $\bar{\beta}$ and average relative pressure drop $\bar{\gamma}$ versus gas mass flux G, as calculated by the model (see Table 1 for values of input parameters).

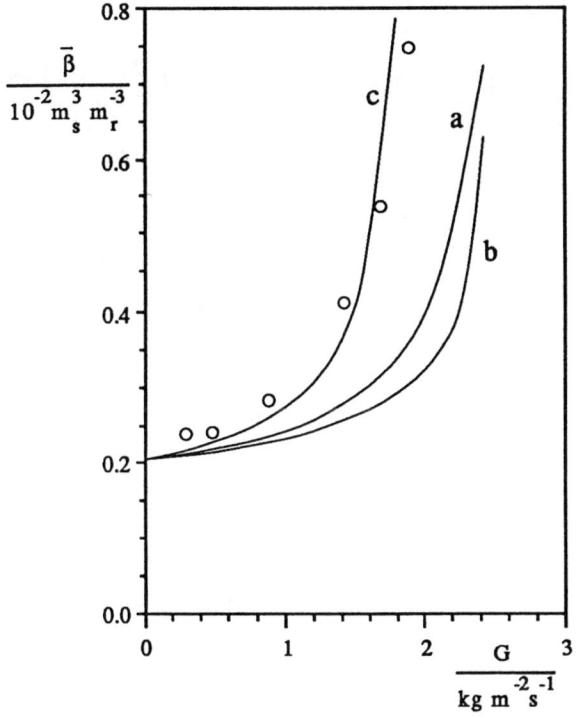
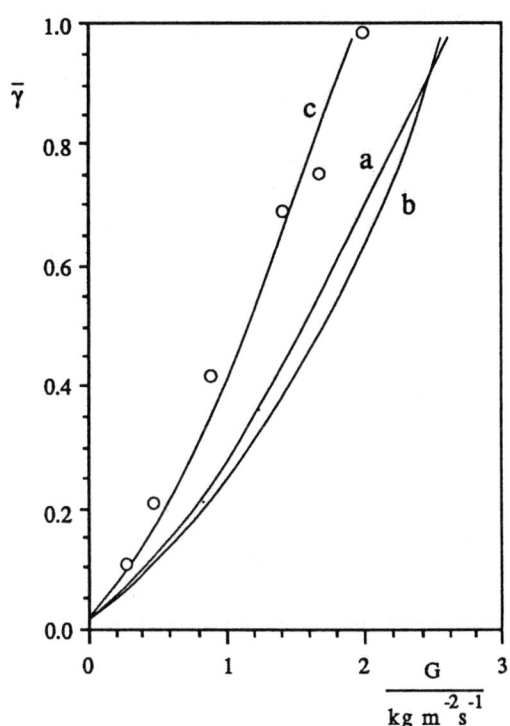

Figure 3. Comparison of different modelling approaches with experimental results (see Table 1 for experimental conditions). In the model computations $u_{so}$ = -0.05 m/s. a) $u_g$ = constant, $\xi$ = 1.0, b) $u_g$ = f(z) according to the axial packing porosity profile (see Equation (24)) and c) $u_g$ = constand, $\xi$ = 1.35.

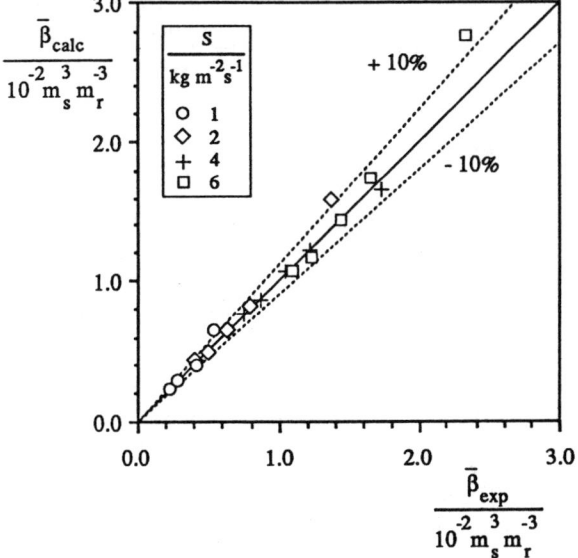
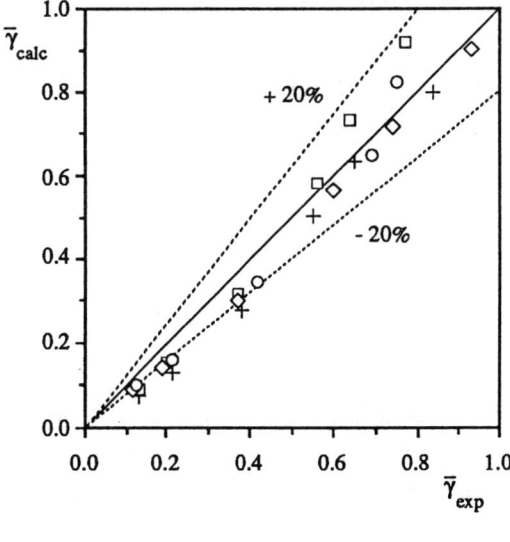

Figure 4. Calculated versus experimental average solids hold-up $\bar{\beta}$ and average relative pressue drop $\bar{\gamma}$ (see Table 1 for experimental conditions).

# COMPUTER SIMULATION OF BUBBLES IN A FLUIDIZED BED

M. Syamlal ■ EG&G W.A.S.C., Inc., Box 880, Morgantown, WV 26507-0880
T.J. O'Brien ■ U.S. Department of Energy, Morgantown Energy Technology Center, Morgantown, WV 26507-0880

A hydrodynamic model, that treats a fluidized medium as a mixture of a gas and a granular phase has been used to study bubble behavior. Bubbles were simulated in fluidized beds of various particle sizes, with and without jets.

The predicted characteristics of bubble formation, coalescence, motion, and eruption at the bed surface are in good qualitative agreement with experimental observations. The computed bubble frequency, bubble volume, rise velocity, wake angle, wake fraction, and pressure profile have been compared with experimental data and simpler theories. The predicted gas and solids mixing have been examined using a new graphical technique and are in agreement with experimental observations.

Bubbles cause many of the fluidization phenomena of practical interest, e.g., gas mixing, solids mixing, and the ejection of solids from the bed. Because of the complexity of multiphase hydrodynamics, most of the theoretical studies on bubbles have been idealized. Numerical simulation of realistic bubbles is now possible because of the availability of high-speed computers and the recently developed numerical codes based on multiphase hydrodynamic equations.

The hydrodynamic modeling of fluidized beds began in the early 1960's. In 1961, Davidson (Davidson and Harrison, 1963) proposed a simple and elegant model of fluidization in terms of the continuity equations for the solids and the gas, and an expression for the relative velocity between the gas and the solids similar to Darcy's Law. Jackson (1963) formulated more general equations of motion for a fluidized bed. Several parallel developments in the 60's and 70's used similar models, but were mainly used to study only the stability of a uniformly fluidized state. In the mid 70's, the energy crisis and the advent of high-speed computers prompted efforts to numerically solve the full set of hydrodynamic equations. Two companies, Systems, Science, and Software and JAYCOR, Inc., and Professor Gidaspow's group at the Illinois Institute of Technology have developed such hydrodynamic models to describe fluidized beds (Gidaspow, 1986).

In this study, a hydrodynamic model is used to simulate bubbles in fluidized beds and the computed bubble characteristics are compared with experimental data. Earlier studies of a similar nature are those of Gidaspow et al. (1983 and 1986), who demonstrated qualitative agreement between simulated and experimental bubbles. This study is more detailed with special emphasis placed on a quantitative comparison with experimental data.

THE HYDRODYNAMIC MODEL

The hydrodynamic representation of a fluidized system consists of balance (mass, momentum, and energy) equations for the gas phase and an interpenetrating granular (solids) phase characterized by a mean diameter, $d_p$, and a density, $\rho_s$. The void fraction e measures the fraction of the volume occupied by the gas. The gas phase continuity equation is

$$\frac{\partial}{\partial t}(\epsilon \rho_g) + \nabla \cdot (\epsilon \rho_g \mathbf{v}_g) = 0 , \qquad (1)$$

and the granular phase continuity equation is

$$\frac{\partial}{\partial t}[(1-\varepsilon)\rho_s] + \nabla \cdot [(1-\varepsilon)\rho_s \mathbf{v}_s] = 0 \ . \quad (2)$$

Momentum balance in the gas phase gives

$$\frac{\partial}{\partial t}(\varepsilon\rho_g \mathbf{v}_g) + \nabla \cdot (\varepsilon\rho_g \mathbf{v}_g \mathbf{v}_g) = -\varepsilon \nabla P$$
$$+ F_{gs}(\mathbf{v}_s - \mathbf{v}_g) + \varepsilon\rho_g \mathbf{g} \ , \quad (3)$$

and, similarly, for the granular phase,

$$\frac{\partial}{\partial t}[(1-\varepsilon)\rho_s \mathbf{v}_s] + \nabla \cdot [(1-\varepsilon)\rho_s \mathbf{v}_s \mathbf{v}_s]$$
$$= -(1-\varepsilon)\nabla P - \nabla P_s + F_{gs}(\mathbf{v}_g - \mathbf{v}_s)$$
$$+ (1-\varepsilon)\rho_s \mathbf{g} \ . \quad (4)$$

The gas-solids drag term, $F_{gs}$, is taken from Syamlal and O'Brien (1987).

$$F_{gs} = \frac{3}{4} C_{DS} \frac{(1-\varepsilon)\varepsilon\rho_g}{V_r^2 d_p} |\mathbf{v}_s - \mathbf{v}_g| \ , \quad (5)$$

where

$$C_{DS} = \left(0.63 + 4.8\sqrt{V_r/Re}\right)^2 \quad (6)$$

$$V_r = 0.5 \ [A - 0.06 Re \quad (7)$$
$$+ \sqrt{(0.0036\ Re^2 + 0.12\ Re(2B-A) + A^2)}\ ] \ ,$$

$$Re = d_p |\mathbf{v}_s - \mathbf{v}_g| \rho_g/\mu_g \ , \quad (8)$$

and

$$A = \varepsilon^{4.14} \ , \quad (9)$$

$$B = \begin{cases} 0.8\varepsilon^{1.28} & \text{for } \varepsilon \leq 0.85 \\ \varepsilon^{2.65} & \text{for } \varepsilon > 0.85 \end{cases} \quad (10)$$

We impose the constraint that $\varepsilon$ should be greater than the void fraction of a randomly packed bed (typically, 0.38). When this constraint is about to be violated in the computations, the void fraction is held constant at this minimum value and the granular pressure $P_s$ is computed as a dependent variable. Thus, $P_s$ takes the place of $\varepsilon$ as an unknown variable. Physically, $P_s$ represents the interparticle reaction forces that maintain the incompressibility constraint and is analogous to the pressure in incompressible fluids.

A transient two-dimensional code based on the above set of equations was developed; the details of the numerical technique can be found in Syamlal (1985, 1987) and in the documentation on K-FIX (Rivard and Torrey, 1977) from which the present code was derived.

## SIMULATION AND DETECTION OF BUBBLES

Several cases of bubble formation and motion were studied numerically. Bubbles were generated either by the injection of a high-speed jet of air (at 20°C) into a minimally fluidized bed or by fluidizing the bed at a velocity twice as large as the minimum fluidization velocity. Table 1 gives a summary of the different conditions of the simulations. Simulations 1 and 3 were performed in a two-dimensional rectangular coordinate system; all others were carried out in an axisymmetric cylindrical coordinate system. Except for Runs 1 and 3, a central jet, 1.0 cm in diameter, was specified; in Run 2, the jet was turned on intermittently for 0.2 seconds. The numerical mesh size was 0.5 cm x 0.5 cm and the time step was 0.0001 s. The initial condition was that of a bed at the minimum fluidization condition with a void fraction of 0.381. The boundary conditions were a specified gas mass-flux from the bottom boundary (the grid flow), a constant pressure of 1 atm at the top

Table 1. Summary of Simulation Conditions

| Simulation Number | $d_p$ (cm) | $\rho_s$ (g/cc) | Bed Dia x Depth (cm x cm) | $U_{mf}$ (cm/s) | Grid Vel. (cm/s) | Jet Vel. (cm/s) |
|---|---|---|---|---|---|---|
| 1 | .020 | 2.48 | 14 x 39 | 6.3 | 12.6 | 0.0 |
| 2 | .055 | 2.48 | 14 x 39 | 29.2 | 29.2 | 450.0 (0.2 s pulses) |
| 3 | .200 | 2.50 | 28 x 39 | 83.5 | 167.0 | 0.0 |
| 4 | .020 | 2.48 | 14 x 39 | 6.3 | 6.3 | Ten velocities |
| 5 | .055 | 2.48 | 14 x 39 | 29.2 | 29.2 | In the range of |
| 6 | .080 | 2.48 | 14 x 39 | 42.1 | 42.1 | 100-1,000 cm/s at |
| 7 | .110 | 2.48 | 14 x 39 | 54.9 | 54.9 | intervals of |
| 8 | .200 | 2.50 | 14 x 39 | 83.5 | 83.5 | 100 cm/s |

boundary, and impermeable side walls with a free-slip condition. In Run 3, the grid flow was present only in the central region, 0 to 7 cm; no flow was present in the region 7 to 14 cm.

Visually, a bubble appears as a region containing almost no particles; its boundary appears to be a discontinuity in the void fraction field. Unfortunately, the numerical technique used smoothes discontinuities. However, based on an evaluation of the inflection point in the void fraction field (the locus of the points where the void fraction is changing most rapidly) and the scatter in the bubble rise velocity versus diameter plots, it has been determined that a void fraction contour of 0.7 adequately represents the outline of a bubble.

## FORMATION OF BUBBLES AT A JET

Runs 4 to 8 were used to determine the effect of particle size on the frequency of bubble formation. The computational results were compared with a simple theory for bubble formation in liquids (Davidson and Harrison, 1963) and was found to deviate systematically for increasing particle size. This is clearly due to the gas leakage from the bubble during its formation. Zhang et al. (1987), have proposed a theory in which the gas is assumed to leak from the top of the bubble at the minimum fluidization velocity $U_{mf}$. Their theory (extended for a spherical bubble) gives the bubble frequency as,

$$f = U_{mf}/\{ a \ln[(a+r_b)/(a-r_b)] - 2r_b \} \quad , \quad (11)$$

where $a = \sqrt{Q/(2\pi U_{mf}\alpha)}$, Q is the volumetric flow rate through the jet, $\alpha$ is the fraction of the hemispherical surface through which the gas leaks, and $r_b$ is the radius of the bubble at the time of detachment from the jet. The bubble radius r is related to s, the distance from the bubble center to the jet inlet, by

$$\frac{d}{dr}\left[ r(a^2 - r^2) \frac{ds}{dr} \right]$$
$$= 4r^5 g/[C(\alpha U_{mf})^2 (a^2 - r^2)] \quad , \quad (12)$$

where C is the virtual mass coefficient. Solving Equation (12) with the initial conditions s = 0 and ds/dr = 0 at r = 0, $r_b$ is obtained as the value of r when r = s.

Figure 1 shows that the trends predicted by the hydrodynamic model and the Zhang et al. (1987), theory are in good agreement; as the particle size (or $U_{mf}$) increases, the bubble volume decreases, and the frequency increases. Taking the physically realistic values of $\alpha = 0.75$ and C = 1.0, the bubble volumes predicted by (12) compare well with those computed by the hydrodynamic model (Figure 1a). From Figure 1b, however, it can be seen that the predicted frequencies are not in agreement. It was not possible to get a good fit in both the plots simultaneously.

## RISE VELOCITY, SHAPE, AND PRESSURE DISTRIBUTION

Figure 2a shows the bubbles in a fluidized bed of small particles (Run 1) at intervals of 0.1 s. The grid flow is steady, uniform, and twice the $U_{mf}$. Numerous tiny bubbles form near the distributor and, subsequently, coalesce into two large bubbles. (The leading bubble was used for all the computations of Simulation 1 presented here since it is free of interference from bubbles in front of it.) As the bubbles rise, the bed expands. When the simulation was continued, however, the bed remained expanded and the bubbling stopped indicating that the fluidization velocity is smaller than the minimum bubbling velocity. Figure 2b shows a bubble generated in Run 2 by pulsing (0.1 to 0.3 s) a high-speed jet. Several bubbles were generated in this manner and were found to be similar. Figure 2c shows bubbles generated in a bed of large particles (Run 3) using a steady, non-uniform grid flow. Similar bubbles formed spontaneously for as long as the simulation was continued. The motion of the bubbles, their coalescence, the eruption at the bed surface, and the consequent motion of the bed surface as seen in Figure 2 are all in good agreement with experimental observations. The predicted rate of bubble expansion is probably a little too large.

The bubble rise velocity, $U_b$, has been given by (Davidson and Harrison 1963) as a function of the diameter, $d_b$, by

$$U_b = 0.711 \sqrt{gd_b} \quad . \tag{13}$$

The bubble center locations and the diameters were computed by drawing the $\varepsilon = 0.7$ void fraction contour and fitting a circle (see the inset in Figure 4a). Using the consecutive bubble center locations at 0.01 s intervals, the rise velocities were computed. The scatter of data points in Figure 3 is because of the uncertainty in the determination of the bubble center locations and the consequent error in the computed rise velocities. The bubble rise velocity, as a function of the bubble diameter, agrees well with Equation (13) for the 0.02 cm particles (Figure 3a) and for the 0.055 cm particles (Figure 3b) for which the experimental data reported by Rowe and Partridge (1965) are also shown. However, for the 0.2 cm particles (Figure 3c), the data does not agree with Equation (13) unless a coefficient smaller than 0.711 is used. Rowe and Partridge also observed that such a coefficient decreases with increasing particle size.

Figure 4 shows the wake angle as a function of the diameter. This dependence on the diameter indicates that the bubble shape changes as the bubble grows and rises through the bed. Close to the distributor, the wake angle is quite large because the distributor plate forces the bubble bottom to be blunt. As the bubble rises, first the wake angle decreases rapidly to almost 0° (nearly a spherical bubble), then it gradually increases, and finally attains a steady-state value as indicated by the large number of overlapping data points. For the 0.02 cm particles (Figure 4a), the initial changes are very erratic, possibly because of the trailing bubble, and the steady-state value of the wake angle is around 110°. For the 0.055 cm particles (Figure 4b), the computed wake angles are compared with the experimental data of Rowe and Partridge (1965). The experimental data, however, are for different bubbles at the same location in the bed and are found to be independent of the bubble diameter. The average value of the experimental data, 135°, is comparable to the steady-state value of 110° from Figures 4a and 4b. From Figure 4c, the steady-state value of wake angle for 0.2 cm particles is 120°, which is again comparable to the experimental data.

The wake fraction, defined as the ratio of the wake volume to the sum of the wake and bubble volumes, was also computed for all the cases. As expected, the variation in the wake fraction was found to correspond to that in the wake angles. The maximum value of the computed wake fraction, approximately 0.2, is comparable to the experimental value of 0.28 (Rowe and Partridge, 1965).

Davidson's theory (Davidson and Harrison, 1963) gives the pressure distribution around an isolated bubble in an unbounded fluidized bed. For a spherical bubble, the normalized pressure distribution along its center line is

$$P_n = \frac{2(P - P_h)}{\rho_s \varepsilon_0 g d_b} = \begin{cases} 1/h^2 & \infty > h \geq 1 \\ h & 1 > h > -1 \\ -1/h^2 & -1 > h \geq -\infty \end{cases} \tag{14}$$

and for a cylindrical bubble

$$P_n = \frac{2(P - P_h)}{\rho_s \varepsilon_0 g d_b} = \begin{cases} 1/h & \infty > h \geq 1 \\ h & 1 > h > -1 \\ -1/h & -1 > h \geq -\infty \end{cases} \quad (15)$$

where $\varepsilon_0$ is the minimum fluidization void fraction and h is the distance from the bubble center, scaled by the bubble radius. $P_h$ is the hydrostatic pressure at a location in the same horizontal plane far enough from the bubble where the pressure field is undisturbed by the bubble. To normalize the computed pressure field, $P_h$ was taken as

$$P_h = P_c - \rho_s \varepsilon_0 g h \quad (16)$$

where $P_c$ is the computed fluid pressure at the center of the bubble.

The distribution of normalized pressure $P_n$ as a function of the scaled distance from the bubble nose (i.e., h - 1) is compared with Davidson's model in Figure 5. Note that when the pressure is plotted in this manner, the constant pressure region inside the bubble is represented by a sloping line (in the range of (h - 1) between 0 and -2 for the theory). The bubble is approximately midway between the distributor and the bed surface; the leftmost simulation point corresponds to the bed surface and the rightmost point corresponds to the distributor. These comparisons are difficult to interpret because of the large size of the bubbles in comparison to the bed dimensions and the proximity of other bubbles and the bed surface. Nevertheless, the comparisons are instructive since they show how the pressure profile is modified due to the nearness of other bubbles and the finite dimensions of the bed.

The pressure profile for the 0.02 cm particles (Figure 5a) differs considerably from Davidson's model (Equation 15) due to the presence of a trailing bubble. Surprisingly, the presence of the second bubble appears to have shifted the constant pressure region (the straight sloping line connecting the two extrema in $P_n$) downwards in the bed. A constant pressure region corresponding to a small bubble near the distributor can be seen at the extreme right-hand side of the figure. The computed bubble in Run 2, being spherical, Equation (14), was used for the comparison in Figure 5b. Within the bubble, the pressure is constant as predicted by Davidson's model. The computed location of the bubble nose, as indicated by the maximum on the left-hand side of the figure, agrees well with the theory and this lends confidence to the void fraction criterion used for identifying bubbles. The normalized pressure is constant above the bubble (left-hand side of the figure) indicating that the pressure field is hydrostatic in nature, but does not become zero, probably, because of the proximity of the bed surface. Below the bubble, the constant pressure region does not extend as far as predicted by Davidson's theory. This is expected since the bubble shape is not spherical as assumed in Davidson's theory. In Figure 5c, for the 0.2 cm particles, since the bubbles are cylindrical, Equation (15) was used for comparisons. At the right side of the figure (close to the distributor), there exists a small region where the pressure is constant, which implies the presence of a bubble. The void fraction contour plot of 0.7 did not show any bubbles in this region. However, from the void fraction data 0.1 seconds later, three small bubbles close to the distributor could be detected. Thus, the small constant pressure region on the right-hand side of Figure 5c appears to indicate evolving bubbles.

GAS AND SOLIDS MOTION NEAR A BUBBLE

Figures 6a and 6b show the gas and granular momenta (product of the velocity and volume fraction of the phases) vector plots. These resemble the velocity vector plots except for the solids within the bubble. The bubble radius is 4.26 cm and the rise velocity is 58.8 cm/s. The plots are for Run 2 at time 0.5 s. At that instant the bubble rise velocity is nearly equal to the interstitial gas velocity. The gas circulation at the sides of the bubble is characteristic of a "fast" bubble. The velocity of the gas at the bubble center is very large. No experimental data or theoretical results are available to verify this. Interestingly,

Davidson's expression for the gas velocity (which is applicable only in the region outside the bubble), if extended, is singular at the bubble center. The notable features of the solids flow are the intense circulation in the wake region and the stagnant region near the distributor.

To study the predicted gas-solids mixing, a computational scheme has been developed to post-process the data generated by the hydrodynamic code. Some "marker particles" are randomly distributed in a specific region of the bed at an initial time. These "particles" are then moved to new locations in every time step using the fluid or solids velocities. The new locations are determined iteratively by assuming that the "particles" move with a velocity that is an average of the velocity at the old location and time and the velocity at the new location and time. A time step of 0.01 s was used for these computations; no convergence problems were encountered.

Rowe and Partridge (1962) placed a layer of colored particles at the bottom of a fluidized bed and demonstrated how a bubble picks up particles in its wake and carries them to the top of the bed. The results of a similar numerical experiment for the 0.055 cm and 0.2 cm particles are shown in Figures 7a and 7b. The particles in a layer at the bottom of the bed are "marked" and are subsequently moved using the granular velocities. The bubble, which forms within this layer, picks up the particles in its wake and transports them to the top of the bed. The particles move in a circulatory manner in the wake region. At the top of the bed, they get distributed at the bed surface. A drift layer originates from the layer at the bottom and slowly moves upwards behind the bubble. Qualitatively, these computational results are in excellent agreement with the experimental observations of Rowe and Partridge (1962). Note that the particles stay close to, but outside, the bubble boundary we have defined, again, confirming $\varepsilon = 0.7$ as the appropriate void fraction criterion to define the bubble boundary.

Figure 8 shows the motion of marker particles initially located randomly inside the bubble and subsequently moved using the gas velocity. The gas mixing plot for Run 1 (Figure 8a) shows that the gas flows out of the bubble and into the trailing bubble. This deviation from Davidson's theory is, clearly, because of the nearness of the trailing bubble. Some of the "marker particles" then reach the sides of the bed and are dragged downwards by the down-flowing solids. Figure 8b (for 0.055 cm particles) shows how the gas flows out from the top of the bubble and flows back in through the sides of the bubble. This circulation pattern is reminiscent of the "cloud" predicted by Davidson's theory for "fast" bubbles and experimentally verified by Rowe (1964). Run 3 is a typical case of a "slow" bubble (Figure 8c). As predicted by Davidson's theory, the gas flows out of the bubble and does not flow back into the bubble.

## CONCLUSIONS

It has been demonstrated that bubble formation, coalescence, motion, and gas and solids mixing caused by the bubbles under various fluidization conditions can be satisfactorily described by a two-phase hydrodynamic model of fluidization.

A study of bubble formation at a jet has shown that, as the particle size (or $U_{mf}$) increases, the bubble volume decreases and frequency increases. These trends are in agreement with a theory of Zhang et al. (1987).

The bubble shape, bubble coalescence phenomena, bubble motion, bubble eruption at the surface, and the dynamics of the bed surface qualitatively agree with experimental observations. Bubble rise velocities were computed and are in good agreement with a formula given by Davidson and Harrison (1963) and the experimental data of Rowe and Partridge (1965). As the bubble rises through the bed, the wake angle changes initially and then attains nearly a steady-state value. The steady-state values of the wake angle are comparable to the experimental data of Rowe and Partridge (1965). The computed wake fractions vary in a manner similar to the wake angle and are in reasonable agreement with the data of Rowe and Partridge (1965).

The computed pressure distributions near the bubble were compared with Davidson's theory (Davidson and Harrison 1963). These comparisons are somewhat tenuous because of the large size of the bubbles in comparison to the bed dimensions and the proximity of other bubbles and the bed surface; these conditions do not conform with those of Davidson's theory. The comparisons clearly show how the pressure profile is modified due to other bubbles and the finite dimensions of the bed. The constant pressure region within the bubble was predicted well.

A graphical method was developed to visualize the mixing of the gas and the solids. The computed flow of gas near the bubble is in qualitative agreement with Davidson's theory and the experimental observations of Rowe (1964). The computed solids mixing mechanism due to the bubble wake is in good agreement with the experimental observations of Rowe and Partridge (1962).

## NOTATION

$d_b$  Bubble Diameter

$d_p$  Particle Diameter

$f$  Bubble Frequency

$F_{gs}$  Gas-Solids Drag

$g$  Gravitational Acceleration

$P$  Gas Pressure

$P_s$  Solids Pressure

$P_n$  Normalized Gas Pressure (Eq. 14 and 15)

$t$  Time

$U_b$  Bubble Rise Velocity

$U_{mf}$  Minimum Fluidization Velocity

$\mathbf{v}_{g,s}$  Velocity Vector (gas or solids)

## Greek Letters

$\varepsilon$  Void Fraction

$\mu_g$  Gas Viscosity

$\rho_{g,s}$  Density (Gas or Solids)

## REFERENCES

Davidson, J.F., and D. Harrison, <u>Fluidized Particles</u>, Cambridge: The University Press, London, 1963.

Gidaspow, D., Applied Mechanics Review, **39**, 1-23 (1986).

Gidaspow, D., Y.C. Seo, and B. Ettehadieh, Chem. Eng. Communications, **22**, 253-272 (1983).

Gidaspow, D., M. Syamlal, and Y.C. Seo, J. of Powder and Bulk Solids Technology, **10**, 19-23 (1986).

Jackson, R., Trans. Inst. Chem. Eng., **41**, 13-28 (1963).

Rivard, W.C., and M.D. Torrey, "K-FIX: A Computer Program for Transient, Two-Dimensional, Two-Fluid Flow." LA-NUREG-6623, April 1977.

Rowe, P.N., Chem. Eng. Progress, **60**, 75-80 (1964).

Rowe, P.N., and B.A. Partridge, in Proc. Symp. on Interaction Between Fluid and Particles, Inst. Chem. Engrs. 135 (1962).

Rowe, P.N., and B.A. Partridge, Trans. Inst. Chem. Eng., **43**, T157-T175 (1965).

Syamlal, M., <u>Multiphase Hydrodynamics of Gas-Solids Flow</u>. Ph.D. Dissertation, Illinois Institute of Technology (1985).

Syamlal, M. and T.J. O'Brien, "A Generalized Drag Correlation for Multiparticle Systems." Unpublished report (1987).

Syamlal, M., "NIMPF: A Computer Code for Non-isothermal Multiparticle Fluidization," EG&G Report, February 1987.

Zhang, X.-R., G.M. Homsy, and W.T. Ropchan, Int. J. Multiphase Flow, **13**, 649-660 (1987).

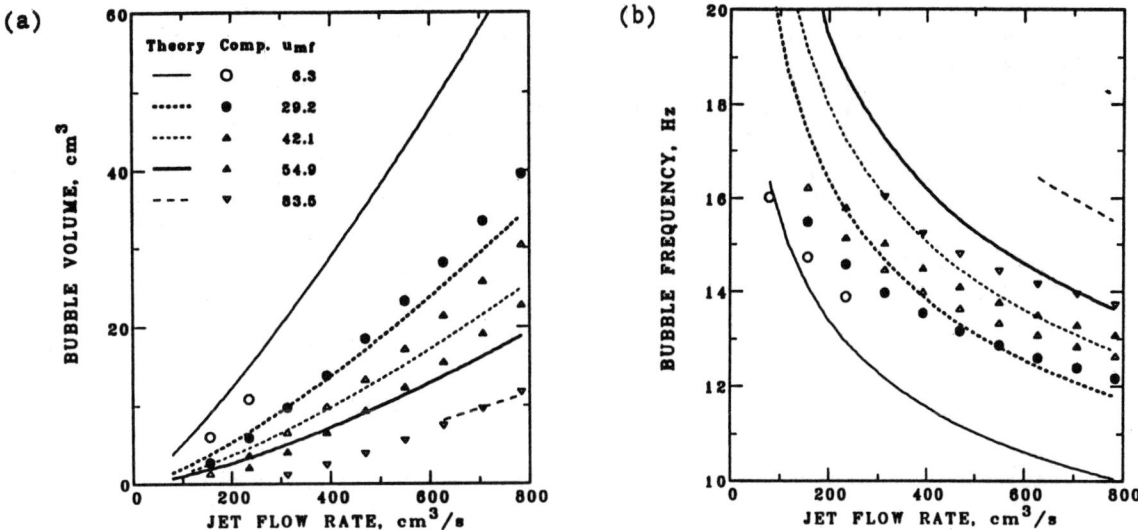

Figure 1. Comparison of computations with the theory of Zhang et al.

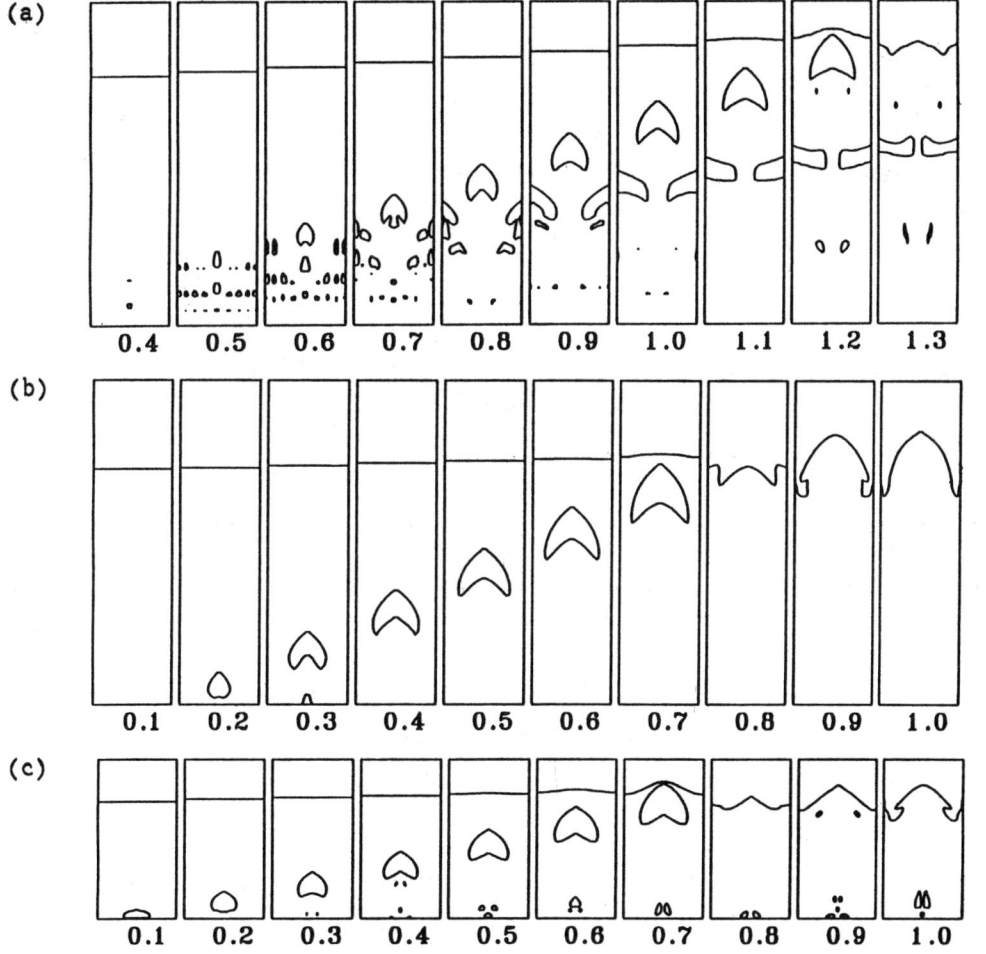

Figure 2. Void fraction contour ($\epsilon=0.7$) plots showing bubble dynamics at intervals of 0.1 s for small particles (a; simulation 1), a pulsed bed (b; simulation 2), and large particles (c; simulation 3)

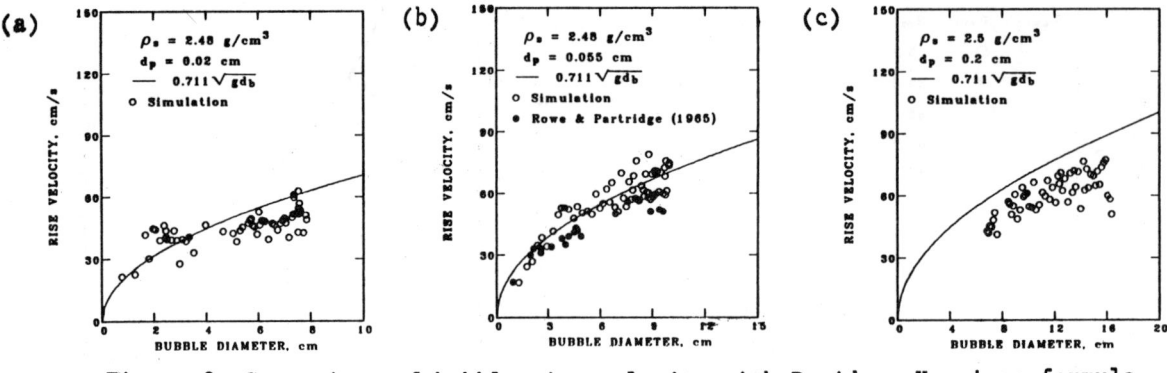
Figure 3. Comparison of bubble rise velocity with Davidson-Harrison formula

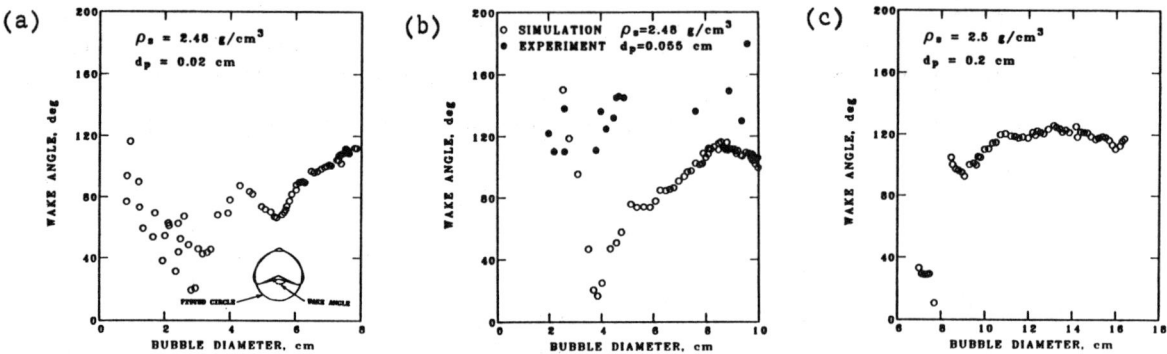
Figure 4. Wake angle change during the growth of a bubble

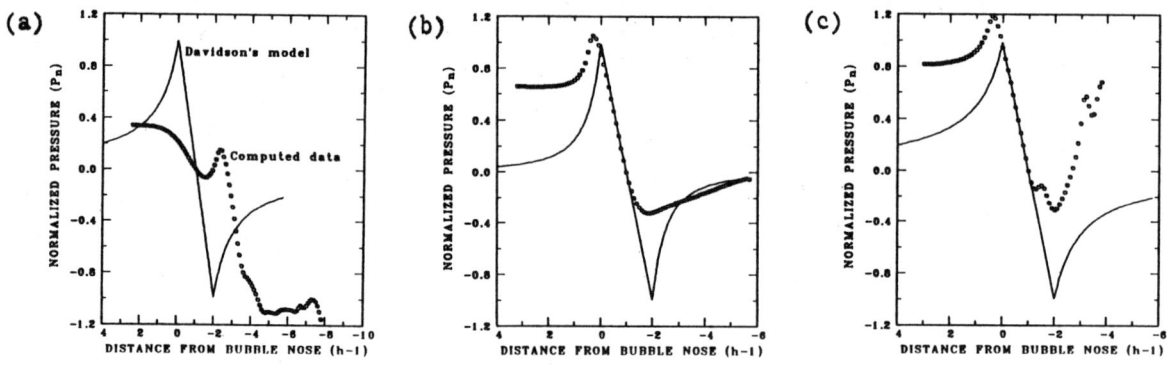
Figure 5. Comparison of pressure distribution with Davidson's bubble model

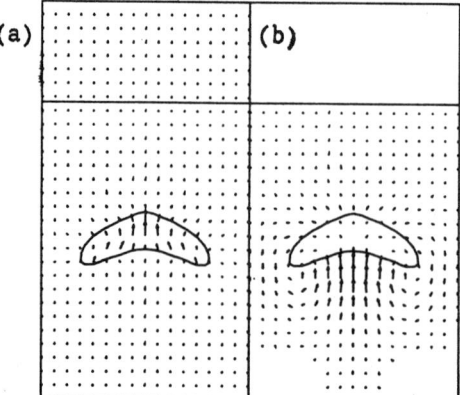
Figure 6. Typical gas (a) and granular (b) momentum vector plots

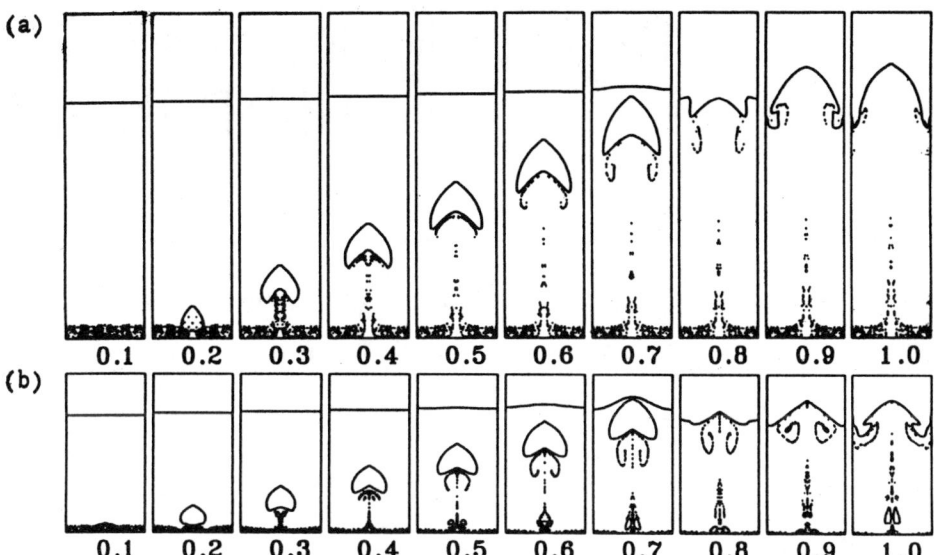

Figure 7. Examples of solids capture in bubble wakes and solids mixing

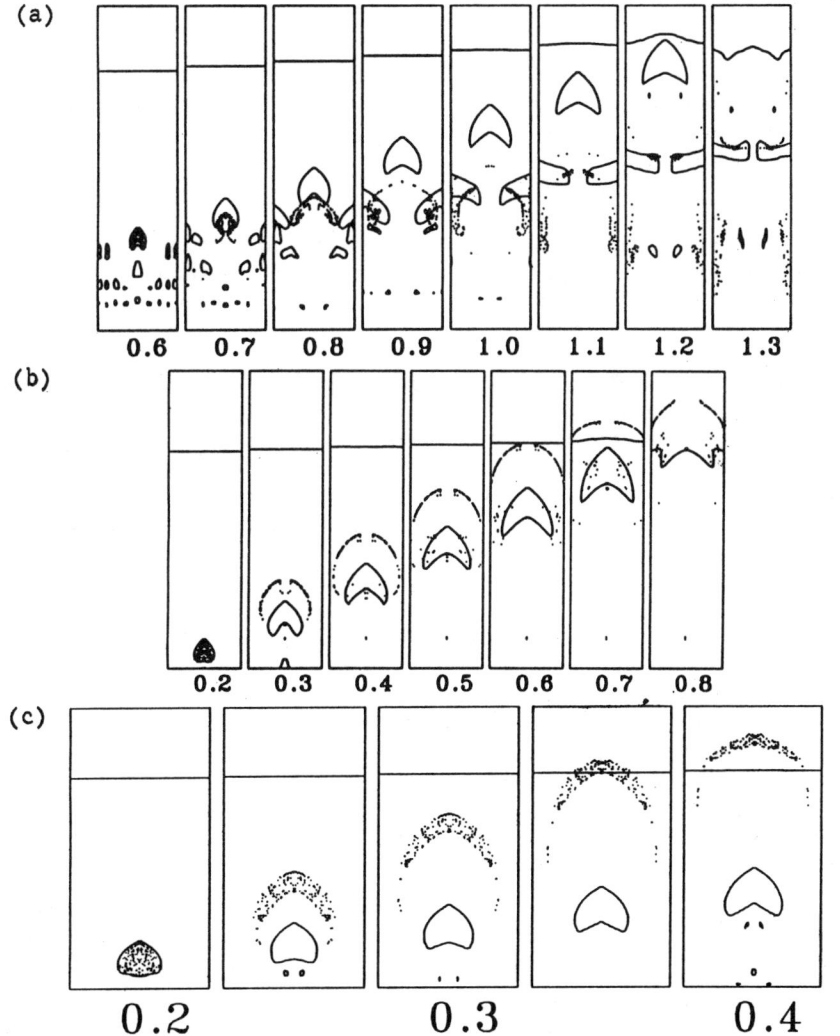

Figure 8. Gas mixing caused by fast (a & b) and slow (c) bubbles

# PARTICLE MIXING IN A CIRCULATING LIQUID FLUIDIZED BED

R. Di Felice, L.G. Gibilaro and S. Rapagna ■ Department of Chemical and Biochemical Engineering, University College London, Torrington Place, London WC1E 7JE UK

P.U. Foscolo ■ Dipartimento di Chimica, Ingegneria Chimica e Materiali, Universita di L'Aquila, 67100 L'Aquila, Italy

A circulating liquid fluidized bed provides a means for obtaining a uniform mixture of different particle species that would, in the conventional non-circulating mode of operation, form separate segregated layers. This phenomenon is examined experimentally for the case of binary particle mixtures and a simple model, that treats the fluid and smaller particles together as a pseudo-fluid in which the larger particles are fluidized, is shown to provide a quantitative explanation of the observed behavior.

Liquid fluidisation receives scant attention in the chemical engineering literature; this is in marked contrast to gas fluidisation which has been widely studied for many years in the universities and industrial research establishments and shows no sign of losing popularity; the growth of biotechnology, with its common emphasis on solid-liquid contacting, has done little to alter the evident imbalance. When it comes to circulatory systems the picture is no different: circulating gas fluidised beds are currently the subject of widespread investigation whereas their liquid counterparts are almost totally ignored. The purpose of this paper is to demonstrate the advantages that can be obtained, with very little equipment modification, by operating liquid fluidised beds in a circulatory mode.

The bed used for the particle mixing studies reported below has been described elsewhere (1): it consists essentially of two vertical tubes connected together to form a closed loop. Fluid and particles are circulated around this loop by means of a marine propellor in the downflow tube. Facilities are available for measuring the axial pressure distribution in both tubes and for suddenly stopping and starting the circulatory flow. This latter feature enables transient bed collapse and expansion measurements to be made which yield the liquid and solid recirculation rates for mono-solid component systems; it has been found that the steady-state solids distribution in this system is completely predictable from the equilibrium relationship, $u(\varepsilon)$, obtained from a standard correlation or, better, measured in a conventional fluidised bed.

In a conventional bed, binary-solid systems tend to form segregated zones with, perhaps, a transitional mixed region between them: bubbles, which promote particle mixing in gas fluidised beds, are normally absent so that the denser, or the larger, particle species segregates to the bottom. The effect of an imposed solids circulation is to continuously transfer the topmost particles to the bottom, thereby tending to promote a uniform distribution throughout the bed.

This phenomenon is now examined on the basis of qualitative observations of the behaviour of binary-solid mixtures fluidised in a circulating bed and the known characteristics of mono-component particle fluidisation: this leads to a very simple model for the expansion of the larger particle species in the presence of smaller circulating particles. Following this, quantitative results of experiments conducted in a circulating bed are compared with the model predictions and found to be in good agreement.

## QUALITATIVE BEHAVIOUR OF A BINARY-SOLID CIRCULATING FLUIDISED BED

Three observed modes of operation are illustrated in Figure 1.

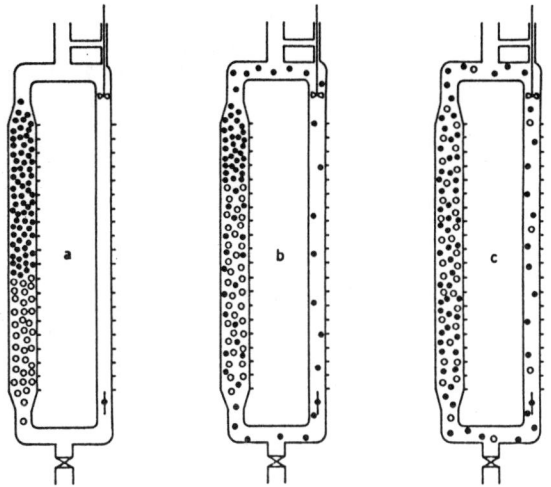

Figure 1. Operating modes for fluidization of strongly segregating particles in a circulating bed.
a) Low fluid velocities: conventional, completely segregated behavior.
b) Intermediate fluid velocities: circulation of smaller particles.
c) High fluid velocities.

At low fluid circulation rates the upflow behaves as a conventional fluidised bed giving rise to two superposed monocomponent zones. At the fluid velocity just above that corresponding to a combined monocomponent zone height equal to the upflow tube height, the topmost species alone circulates: this gives rise to a bottom binary-solid zone in the upflow tube consisting of the static bottom segregating component being percolated by the upper segregating component, and a zone above of the pure circulating species. At yet higher fluid velocities, the bottom segregating component starts to circulate giving rise to a completely uniform mixture in the upflow tube.

The first critical fluid velocity, $u_{C1}$, at which the upper segregating species alone starts to circulate, is immediately determinable from the mass of the two components and their individual expansion characteristics which are usually expressed in the form:

$$u = f(\varepsilon), \quad (1)$$

where u is superficial fluid velocity in a conventional equilibrium bed. A conservative estimate of the second critical velocity, $u_{C2}$, at which both species start to circulate, is simply the fluid velocity required to expand the bottom segregating species alone, in the absence of the second component, to the top of the upflow tube: in fact, due to the presence of the other component, this overestimates the fluid velocity required to give a completely uniform distribution in the upflow tube. We now consider how the quantitative effect of the smaller circulating component on the expansion of the larger particle zone can be estimated.

## THE PSEUDO-FLUID MODEL

The situation depicted in Figure 1(b), of an equilibrium bed of the larger particles being percolated by both the fluid and the smaller particles suggests that it can be treated as an equivalent monocomponent system: we thus consider the case of the larger particles fluidised by a pseudo-fluid having physical properties appropriate to the smaller particle/fluid suspension; once these properties have been identified the monocomponent correlations, that relate to explicit formulations of Equation (1), may be applied to the estimation of the expansion characteristics of the larger particle bed. The problem thus reduces to defining the density, viscosity and volumetric flux of the pseudo-fluid.

The density poses no problem: it must be evaluated as the mean value for the smaller particle/fluid suspension:

$$\bar{\rho} = \frac{C_S \rho_S + \varepsilon \rho}{C_S + \varepsilon} \quad (2)$$

The contribution to the total mean pressure gradient in the bed arising from a fluid of density $\bar{\rho}$ is clearly identical to that produced by the pure fluid with suspended smaller particles.

The appropriate value to take for the pseudo-fluid viscosity is less certain; however, as the larger particles are in contact with pure fluid a first approximation could be the pure fluid viscosity:

$$\bar{\mu} = \mu . \quad (3)$$

This assumption may require reconsideration in the light of experimental observations.

The appropriate volumetric flux of the pseudo-fluid would appear to be the sum of the volumetric fluxes of pure fluid and smaller particles:

$$\bar{u} = u_f . \varepsilon + u_s C_S . \quad (4)$$

However, for the experimental conditions considered below, we have that both the fluid velocity, $u_f$, is much greater than the smaller particle velocity, $u_s$, and the volumetric fraction of pure fluid, $\varepsilon$, is much greater than that of the smaller particles, $C_S$, so that:

$$\bar{u} \simeq u_f \varepsilon = u . \quad (5)$$

This is a convenient approximation because, together with the assumption of Equation (3), it renders predictions of the larger particle expansion characteristics independent of the smaller particle circulation rate.

## EXPERIMENTAL

Experiments on binary-solid particle mixtures fluidised by water were performed in the circulating bed described briefly above and in more detail in (1).

In all cases the bottom segregating component consisted of 5 mm acetate spheres; three different top segregating components were utilised: 4 mm and 2 mm acetate and 420 μm lead glass spheres.

The experiments consisted of charging the system with a fixed amount of the 5 mm acetate spheres and observing the variation in height of the bottom segregated zone with fluid velocity for various additions of one of the smaller particle species: in all, the results of 16 different overall bed compositions are reported and these are specified in Table 1.

In each case the bed behaved qualitatively exactly as described above and depicted in Figure 1: complete segregation of the two components was observed for fluid velocities below that required to expand the two monocomponent zones to the top of the upflow tube, $u_{C1}$; thereafter the bottom binary zone

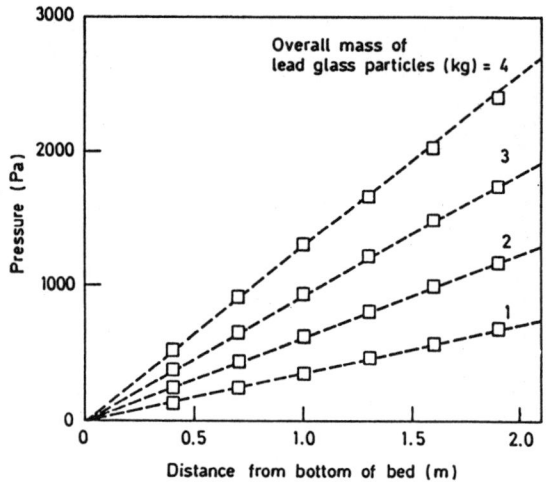

Figure 2. Axial pressure distribution in the up-flow tube for the fully expanded 5 mm acetate particle bed (of mass 1.00 kg) in the presence of 420 μm lead glass particles.

expanded in a regular manner with increasing circulation until it reached the top of the upflow tube at the second critical velocity, $u_{C2}$. The bottom zone always consisted of a uniform mixture of particles, there being no discernable concentration gradients along its entire length: this was confirmed from axial pressure measurements made at intervals along the column and illustrated for examples of just fully expanded systems (at the second critical fluid velocity, $u_{C2}$) in Figure 2.

These pressure distribution measurements enable the composition of the bottom binary zone to be evaluated: the larger particle concentration, $C_\ell$, is known directly from the height of the bottom zone so that the force balance relationship,

$$\frac{dp}{dz} = [\rho + C_s(\rho_s - \rho) + C_\ell(\rho_\ell - \rho)] \quad (6)$$

yields the smaller particle concentration, $C_s$ (it having been previously established that wall friction contributes negligibly to the pressure gradient). Values of $C_s$, for all the fully expanded beds, are reported as points in Figure 3 where they are compared with predictions based on the pseudo-fluid model obtained as described below.

Fluid velocities were estimated by measuring the falling velocities of individual larger particles introduced to the top of the downflow tube; circulating particle concentrations in this tube were always very low so that the slip velocity for the introduced particle corresponded to $u_t$ - its unhindered

Table 1. Particle properties and overall system compositions.

| SEGREGATING COMPONENT | | | | | | | | Symbols |
|---|---|---|---|---|---|---|---|---|
| BOTTOM | | | | TOP | | | | |
| Material | Density (kg/m³) | Size (mm) | Weight (kg) | Material | Density (kg/m³) | Size (mm) | Weight (kg) | |
| Acetate | 1270 | 5.00 | 4 | Acetate | 1270 | 2.00 | 0.5, 1, 1.5, 2 | ▼ |
| | | | 1 | Lead glass | 2900 | 0.42 | 1, 2, 3, 4 | □ |
| | | | 2 | Lead glass | 2900 | 0.42 | 0.5, 1, 1.5, 2 | ◆ |
| | | | 4 | Acetate | 1270 | 4.00 | 0.5, 1, 1.5, 2 | △ |

settling velocity in stagnant fluid; $u_t$ was measured directly in the downflow tube thereby enabling fluid circulation rates to be determined.

## PREDICTION OF THE SECOND CRITICAL VELOCITY

### An upper bound on $u_{C2}$

A conservative estimate of $u_{C2}$ has been referred to above: this is obtained by disregarding completely the effect of the smaller particles in the bottom zone and applying an explicit formulation of the equilibrium relationship, Equation (1), evaluated for the fully expanded monocomponent bed of larger particles:

$$u_{C2} = f(1 - M_\ell/AL\rho_\ell) \qquad (7)$$

The explicit form employed was:

$$u = Ku_t\varepsilon^n. \qquad (8)$$

In applying Equation (8), K is usually taken to be unity in which case $u(\varepsilon)$ can be represented by a single straight line on logarithmic coordinates; however, it has long been known (2) that for high terminal Reynolds number systems (with which the present study is concerned) there is a break in this linear relationship that must be accommodated by two log-linear segments having the form of Equation (8). A recent study of monocomponent bed expansion, that utilised the same high precision acetate spheres used here (3) provided general correlations for the parameters of Equation (8) which have been used for the predictions reported in the present study: the upper bound for $u_{C2}$ corresponds to the values shown in Figure 3 for $C_s$ equal to zero.

### The pseudo-fluid model predictions of $u_{C2}$

The role of the circulating smaller particles in promoting a more rapid expansion of the bottom zone may be examined by substituting the pseudo-fluid properties into the monocomponent correlations for the parameters of Equation (8): for reasons described above this involves simply replacing the fluid density, $\rho$, by that of the pseudo-fluid, $\bar{\rho}$, Equation (2).

The results of these simulations are shown by the continuous lines in Figure 3.

## DISCUSSION

A circulating liquid fluidised bed has been shown to provide a perfectly homogeneous blend of particle species that, in a conventional non-circulating bed, would exhibit complete segregation. Although the reported study involves solely binary spherical particle systems, the qualitative generalisation, to 'real' process environments, involving perhaps many solid components of ill defined shape, would appear likely and certainly worthy of further study.

The results reported in Figure 3 are of interest in a number of respects. In the first place they vindicate the very simple pseudo-fluid hypothesis which in all cases provides a close agreement with the observed behaviour. Other similar treatments of a solid-fluid suspension have been described by previous workers and applied to related problems: thus, for example, Barnea and Mizrahi (4) consider a single particle, in a monocomponent fluidised suspension, to be surrounded by a pseudo-fluid that has the mean density of the suspension and a viscosity that is a function of particle concentration; and Selim et al. (5), in a study of multisize particle sedimentation, use precisely the pseudo-fluid properties adopted here, $\mu$ and $\bar{\rho}$, to account for the hindering effect of the smaller particles, but then treat the equilibrium relationship for the larger particles somewhat differently by retaining the actual fluid voidage in Equation (8) rather than that based on the pseudo-fluid which incorporates the holdup of the smaller particles.

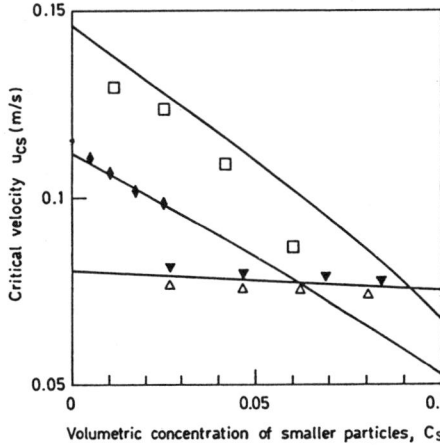

Figure 3. Effect of smaller particle concentration in up-flow tube on second critical velocity.
Experimental results (points: symbols as given in Table I) compared with model predictions (continuous lines).

Figure 3 illustrates very convincingly the dominant role of the pressure field (or effective buoyancy) on the expansion of the larger particles: the two cases representing the effect of glass particles on the second critical velocity show that an increase in their volumetric concentration causes a marked reduction in $u_{C2}$; for the other two cases, where the smaller particles were of acetate, very little change in critical velocity was observed so that the 'upper bound' value (for $C_S = 0$) adequately predicted the behaviour in all cases. This is clearly because glass particles significantly influence the mean density of the surroundings whereas the acetate particles of density much closer to that of the fluid have little effect.

## NOTATION

| | |
|---|---|
| A | area of upflow tube, $m^2$ |
| $C_\ell$, $C_S$ | volumetric concentration of larger and smaller particles |
| f | equilibrium relationship, Equation (1), $ms^{-1}$ |
| K | parameter in Equation (8) |
| L | length of upflow and downflow tubes, m |
| $M_\ell$ | total mass of larger particles, kg |
| p | pressure, $Nm^{-2}$ |
| $u$, $\bar{u}$ | superficial velocity of fluid and pseudo-fluid, $ms^{-1}$ |
| $u_{C1}$, $u_{C2}$ | 1st and 2nd critical fluid superficial velocities, $ms^{-1}$ |
| $u_f$, $u_s$ | velocities of fluid and smaller particles, $ms^{-1}$ |
| $u_t$ | terminal unhindered settling velocity of a larger particle, $ms^{-1}$ |
| z | distance variable, m |
| $\varepsilon$ | void fraction, volumetric fraction of fluid |
| $\mu$, $\bar{\mu}$ | viscosity of fluid and pseudo-fluid, $Nsm^{-2}$ |
| $\rho$, $\bar{\rho}$ | density of fluid and pseudo-fluid, $kg\ m^{-3}$ |
| $\rho_S$, $\rho_\ell$ | density of smaller and larger particles, $kg\ m^{-3}$ |

## LITERATURE CITED

1. Gibilaro, L.G., R. Di Felice and P.U. Foscolo, Chem. Engng Sci., 43, 2901 (1988).

2. Garside, J. and M.R. Al-Dibouni, Ind. Eng. Chem. Process Des. Dev., 16, 206 (1977).

3. Rapagna, S., R. Di Felice, L.G. Gibilaro and P.U. Foscolo, Chem. Engng Communications, In press.

4. Barnea, E. and J. Mizrahi, Chem. Engng Journal, 5, 171 (1973).

5. Selim, M.S., A.C. Kothari and R.M.C. Turian, AIChE Journal, 29, 1029 (1983).

# EFFECT OF OPERATING TEMPERATURE AND PRESSURE ON THE TRANSITION FROM BUBBLING TO TURBULENT FLUIDIZATION

P. Cai, S.P. Chen, Y, Jin, Z.Q. Yu and Z.W. Wang ■ Department of Chemical Engineering, Tsinghua University, Beijing, 100084, P.R. China

Experiments have been carried out to investigate the effects of operating temperature (50 to 500 °C) and pressure ($1 \times 10^5$ to $8 \times 10^5$ Pa) on the transition from bubbling to turbulent fluidization, based on computer analysis of the pressure fluctuation in dense phase. In order to reveal the influence of particle properties on the transition process, eight kinds of particles were employed. A empirical correlation, which takes into account the effects of temperature and pressure, is recommended for the prediction of transition velocity.

## INTRODUCTION

Turbulent fluidization is a operating regime between bubbling and fast fluidization. Due to its high contact efficiency between the gas and particles, this regime could be shown to have some advantages over the other regimes in terms of the stable operation, the high productivity and easy of scaling up, as well as the high transfer efficiency. So it is superior for some industrial catalytic reactors to be operated in this regime.

It has been found that the basic features of the fluidized bed will change with the raising of bed temperature and pressure (1 - 7). Therefore, the influence of temperature and pressure can not be ignored for the operation of commercial fluidized bed reactors; and the studies on effects of temperature and pressure on the transition from bubbling to turbulent fluidization are essential for the design and operation of fluidized bed reactors.

## EXPERIMENTAL APPARATUS AND METHODS

Two sets of experimental apparatus were employed. The first one for the temperature effect study, which was 150mm in diameter and 3800mm in height, operated at ambient pressure (as shown in Figure 1). The study on pressure effect was conducted in another one, which was 284mm in diameter and 2500mm in height, and operated at room temperature. In order to take into account the influences of particle properties, eight kinds of particles were employed (belonging to A and B groups in Geldart's classification). The physical properties of the particles were listed in Table 1. The operating temperature ranged from 50°C to 450°C, and the variation of operating pressure was between $1 \times 10^5$ Pa and $8 \times 10^5$ Pa.

Table 1. The Properties of Experimental Particle

| PARTICLES | $d_p$ μm | $\rho_p$ kg/m³ | $\varepsilon_0$ | $\rho_B$ kg/m³ |
|---|---|---|---|---|
| Silicagel A | 476 | 834 | 0.349 | 543 |
| Silicagel B | 280 | 706 | 0.351 | 459 |
| Silicagel E | 165 | 711 | 0.409 | 420 |
| Silicagel F | 1057 | 844 | 0.425 | 485 |
| P-Y FCC | 65 | 1172 | 0.511 | 573 |
| Y-7 FCC | 53 | 1667 | 0.475 | 876 |
| Resin | 566 | 1330 | 0.371 | 836 |
| Quartz sand | 98 | 2580 | 0.473 | 1359 |

Based on the analysis of pressure fluctuation

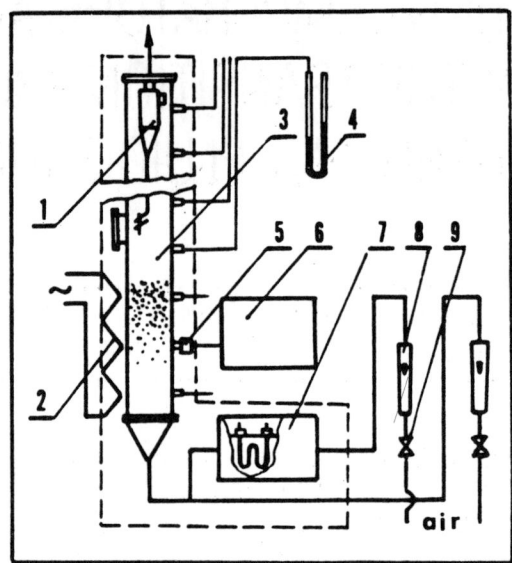

1 – cyclone  
2 – heater  
3 – fluidized bed  
4 – U – manometer  
5 – pressure transducer  
6 – data analysis system  
7 – heating chamber  
8 – rotameter  
9 – valve  

Figure 1. Experimental Apparatus for the Temperature Effect

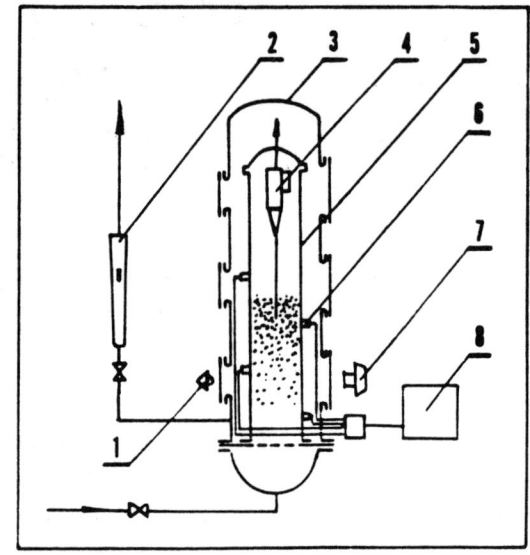

1 – lamp  
2 – rotameter  
3 – pressurized shell  
4 – cyclone  
5 – fluidized bed  
6 – pressure transducer  
7 – camera  
8 – data analysis system  

Figure 2. Experimental Apparatus for the Pressure Effect

in the fluidized bed, Y. Jin et al. (8) proposed the following correlation to calculate the transition velocity $u_c$ at room temperature and ambient pressure,

$$\frac{u_c}{\sqrt{g\,d_p}} = \left[ K \left( \frac{\rho_p - \rho_f}{\rho_f} \right) \left( \frac{D}{d_p} \right) \right]^{0.27} \quad (1)$$

Considering the influence of bed diameters, P. Cai et al. (9) obtained the following modified equation for free fluidized bed

$$\frac{u_c}{\sqrt{g\,d_p}} = \left( \frac{0.211}{D^{0.27}} + \frac{2.42 \times 10^{-3}}{D^{1.27}} \right) \left[ \left( \frac{\rho_p - \rho_f}{\rho_f} \right) \left( \frac{D}{d_p} \right) \right]^{0.27} \quad (2)$$

In this study, the same method was applied for the determination of transition velocity $u_c$.

For a given operation condition with low gas velocity, the bed will be found at bubbling regime. By increasing the gas velocity, the bubbles will grow up, then the coalesce of the bubbles will be more significant in comparison with their break up. As a result, the magnitude of pressure fluctuation signals will increase rapidly, and the calculated standard deviation S will rise too (as shown in Figure 3).

$$S = \sqrt{\frac{1}{T} \int_0^T (p(t) - \bar{p})^2 dt} \quad (3)$$

When the gas velocity exceeds the critical point $u_c$, due to the highly increased break up frequency of the bubbles, the bed is transformed from bubbling to turbulent fluidization. At this time, it could be noted that the standard deviation of signals are being reduced gradually, and a peak appears in the curve of standard deviation versus gas velocity. Obviously, this peak is corresponding to the transition point. Therefore, the transition velocity could be obtained by such experimental approach.

## RESULTS AND DISCUSSIONS

### Effects of Operating Temperature on Transition Velocity

The standard deviation of pressure fluctua-

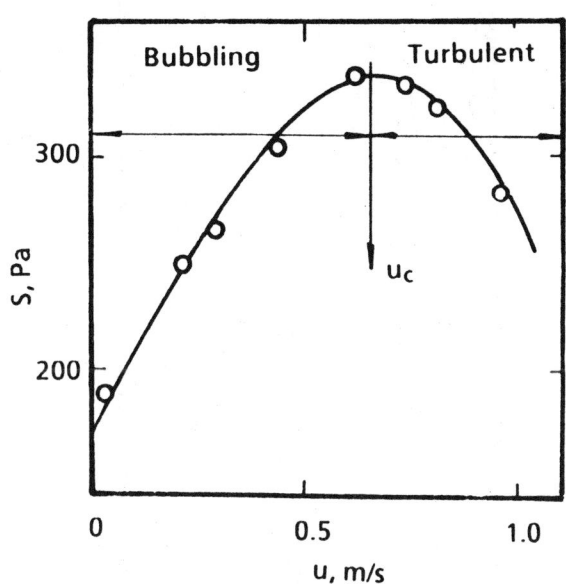

Figure 3. Typical Experimental Data

tion S was plotted against gas velocity u at different temperatures (50°C, 200°C and 450°C) in Figure 4 (a). It can be seen that the higher the operating temperature, the smaller the value of S and $S_{max}$. This is consistent with the reported results (3, 4), i. e., the bubble size decreased with increasing operating temperature.

The experimental results also show that the transition velocity $u_c$ increases with the increase of bed temperature; i. e., the initial transition from bubbling to turbulent fluidization is delayed. Obviously, the higher the bed temperature, the higher the gas viscosity and lower the gas density. It is clear that the effects of the operating temperature on the behavior of gas-solid system are achieved by means of changing the gas viscosity and density. In order to investigate the effect of gas viscosity on the magnitude of pressure fluctuation and the transition process, the experimental data, which mentioned above, were re-plotted in Figure 4(b) to show the relationship between gas flow rate G and standard deviation S. It can be seen that the transition will occur at lower $G_c$ for higher operating temperature. This may imply that the delay of transition at higher temperature is mainly resulted by the decrease of gas density. This conclusion is proved to be true also by the further study on the pressure effect.

In order to study the effect of particle size under high temperature, several kinds of silicagel particles with different size were used. As shown in Figure 5, for both kinds of particles, the standard deviation S will decrease and transition velocity $u_c$ will increase with increasing operating temperature. In the experimental temperature range (50 to 450°C), by making comparison between the results obtained at same temperature level, it was shown that both the standard deviation S and transition velocity $u_c$ are higher for coarse particles.

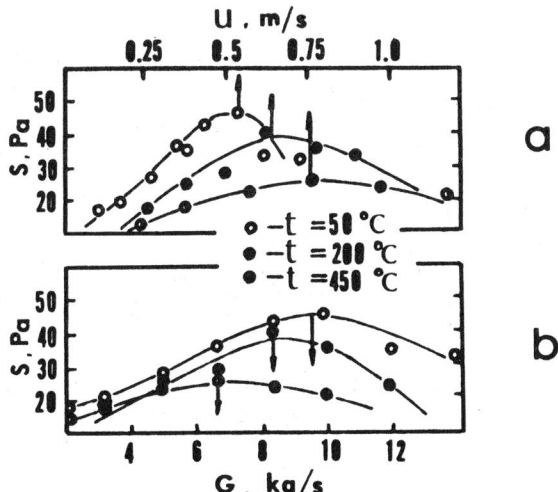

Figure 4. Effects of Temperature on Flow Regime Transition

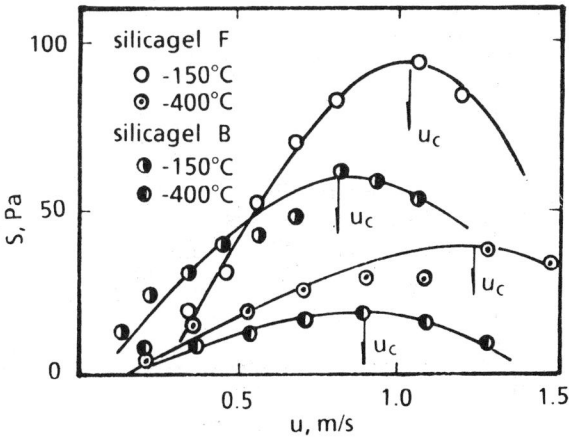

Figure 5. Effect of Particle Size on Flow Regime Transition

Similarly, two kinds of FCC powders with approximate size and different densities were employed to investigate the effect of particle density on transition velocity. Figure 6 shows the similar trend as the influence of particle size discussed above.

The signals of pressure fluctuation obtained under transition velocity $u_c$ for different temperatures were shown in Figure 7. It can be seen that the values of $S_{max}$ decrease continuously with the temperature rising from 50°C to 450°C. By plotting $S_{max}$ against operating temperature in Figure 8, it can be seen that for coarse particles $S_{max}$ decreases considerably with the increase of operating temperature. But for fine powders, the variation is less significant.

Effect of Operating Pressure on Transition Velocity

Figure 9 shows the relationship between the standard deviation and operating gas velocity for silicagel A, B and Y-7 FCC particles under different operating pressures. It also indicates that the maximum point $u_c$ will be lowered with the increase of operating pressure. In other words, the transition will occur in advance with the increase of operating pressure (as shown in Figure 10).

At a given gas velocity and with the increas-

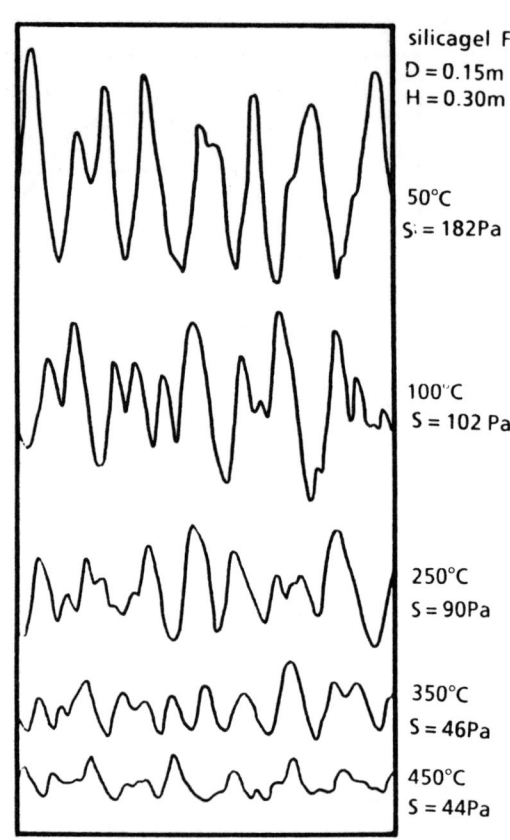

Figure 7. Signals of Pressure Fluctuation at the Transition Velocity

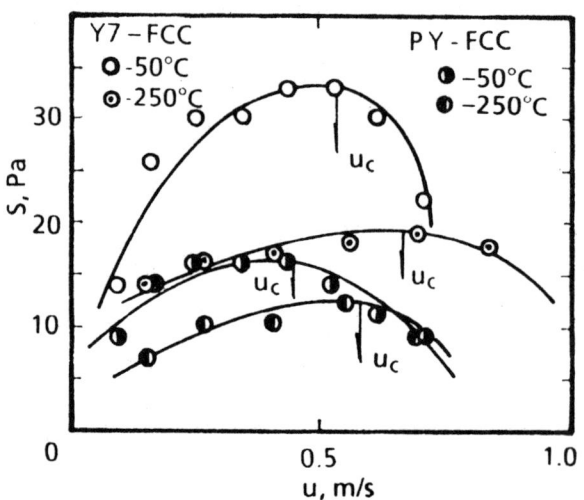

Figure 6. Effect of Particle Density on Flow Regime Transition

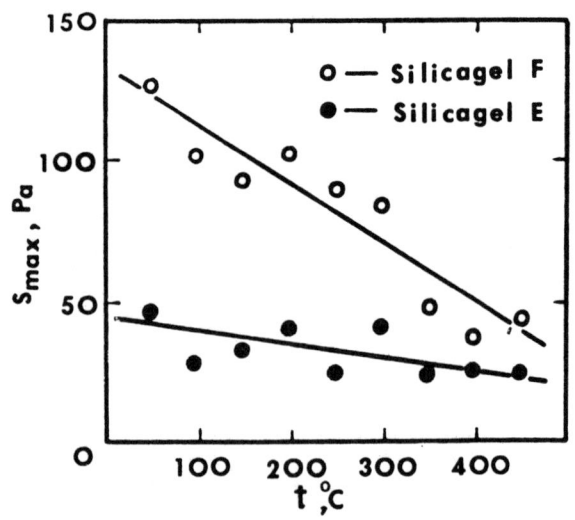

Figure 8. Relationship between $S_{max}$ and Operating Temperature

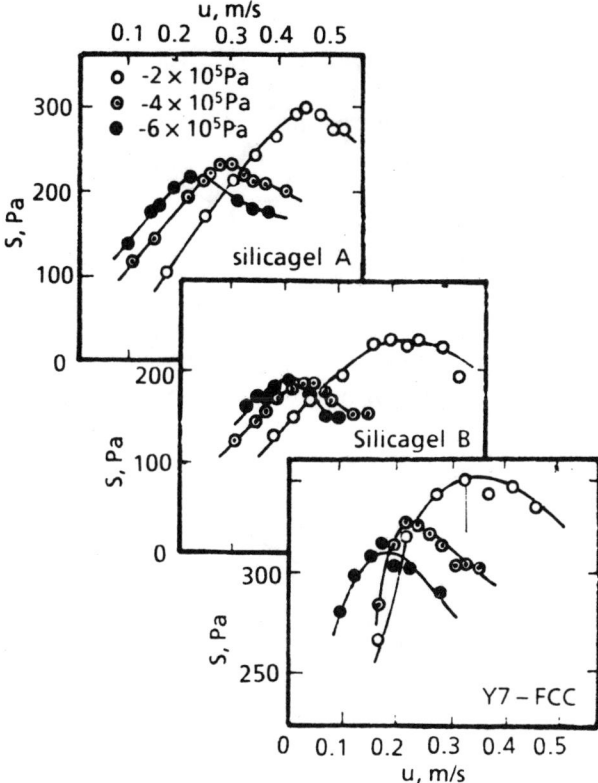

Figure 9. Relationship between S and Gas Velocity

ing of operation pressure, the standard deviation will increase to reach a peak point and then decreases gradually (Figure 11). The higher the gas velocity, the higher the peak point. The corresponding pressure can be defined as transition pressure $P_c$. This is consistent with Rowe et al.'s (6) experimental results for bubble size and frequency in pressurized bed.

From above analysis, it can be seen that the standard deviation S is a function of gas velocity u, operating pressure P and particle properties. For a given kind of particles, a three dimensional diagram $S = f(u, P)$ could be drawn as Figure 12. Therefore the regularity of variation of transition with operating pressure is expressed more clearly.

## The Correlation of Experimental Data

On the basis of the data measured under different operation pressure and temperature for eight kinds of particles, the following correlation

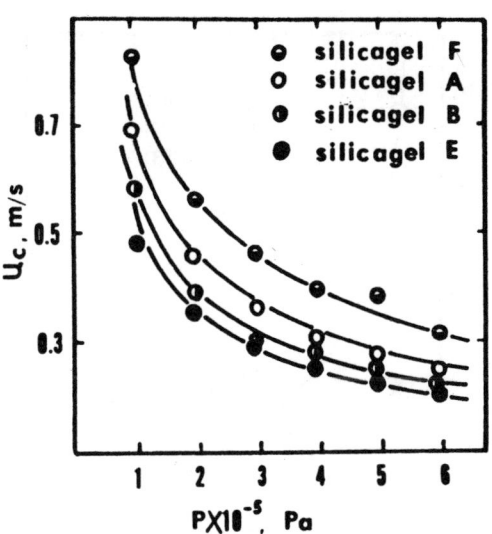

(a) Silicalgel Particles

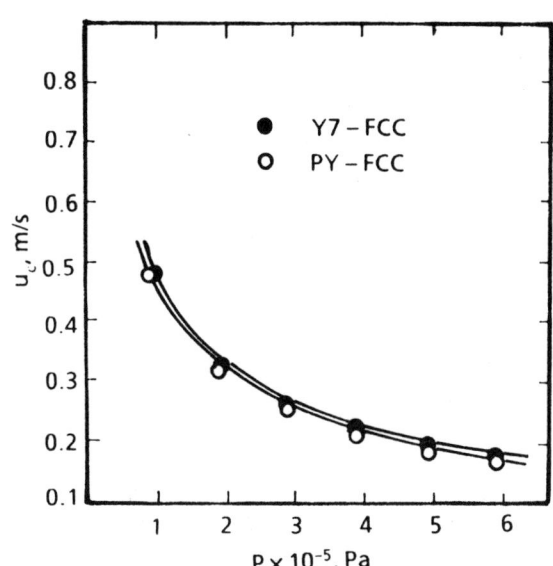

(b) FCC Powders

Figure 10. Effect of Operating Pressure on Transition Velocity

was recommended for the prediction of transition velocity $u_c$:

$$\frac{u_c}{\sqrt{g\,d_p}} = \left(\frac{\mu_{f20}}{\mu_f}\right)^{0.2} \left[K\left(\frac{\rho_{f20}}{\rho_f}\right)\left(\frac{\rho_p - \rho_f}{\rho_f}\right)\left(\frac{D_f}{d_p}\right)\right]^{0.27} \quad (4)$$

where

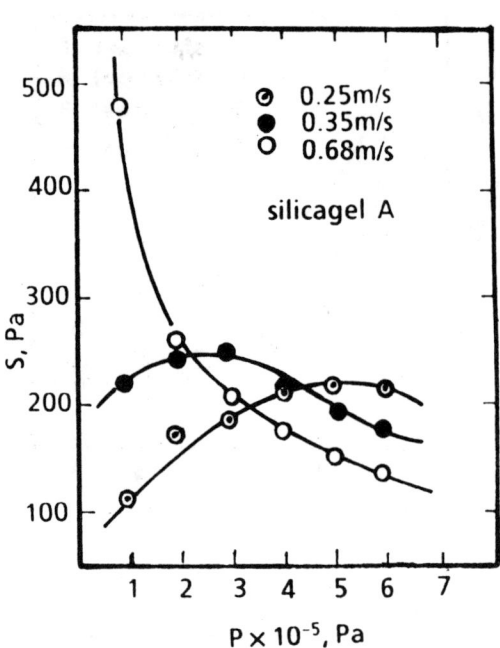

Figure 11. Relationship between S and Operating Pressure

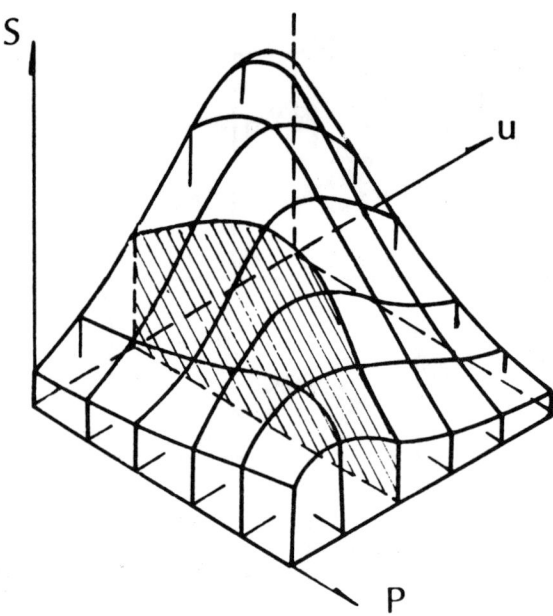

Figure 12. The Three Dimensional Diagram $S = f(u, P)$

$$D_f = D, \quad K = \left( \frac{0.211}{D^{0.27}} + \frac{2.42 \times 10^{-3}}{D^{1.27}} \right)^{\frac{1}{0.27}}$$

free bed

$$KD_f = (1.64 \text{ to } 2.32) \times 10^{-3}$$

bed with vertical tubes

$$KD_f = 3.42 \times 10^{-4}$$

bed with pagoda type internal baffles (from Jin et al. (10))

$$KD_f = 4.69 \times 10^{-3}$$

2-dimensional bed (12 x 300mm)

The particle properties, bed diameter, operating pressure and temperature have been taken into account in this equation.

The average error for above correlation is about ±5%.

## CONCLUSIONS

1. With the increase of operating temperature, the transition from bubbling to turbulent fluidization will be delayed; but with increasing operating pressure, the transition will occur in advance.

2. The influence of temperature and pressure will be more significant for coarse and heavy particles.

3. There exists a pressure transition point $P_c$ and the value of $P_c$ will decrease with increasing gas velocity.

## NOTATION

### Roman Letters

| | |
|---|---|
| D | diameter of fluidized bed [m] |
| $D_f$ | equivalent diameter of fluidized bed [m] |
| $d_p$ | mean size of particles [m] |
| f | frequency of pressure fluctuation [HZ] |
| g | gravitational acceleration velocity [m/s$^2$] |
| G | mass flow rate of the gas [kg/s] |
| K | constant |
| p(t) | pressure fluctuation signal [Pa] |
| $\bar{p}$ | the average value of p(t) [Pa] |
| P | operating pressure [Pa] |
| $P_c$ | transition pressure from bubbling to turbulent fluidization [Pa] |

S standard deviation of pressure fluctuation [Pa]
$S_{max}$ S at transition velocity [Pa]
t time [s]
T period of sampling [s]
u superficial gas velocity [m/s]
$u_c$ transition velocity from bubbling to turbulent [m/s]

Greek Letters

$\rho_p$ particle density [kg/m$^3$]
$\rho_f$ gas density [kg/m$^3$]
$\rho_{f20}$ gas density at 20°C and $1 \times 10^5$ Pa [kg/m$^3$]
$\mu_f$ gas viscosity [Pa·s]
$\mu_{f20}$ gas viscosity at 20°C [Pa·s]

## LITERATURE CITED

1. Nakamura, M., Y. Namada, and S. Toyama, Canadian J. Chem. Eng., 63, 8 (1985).
2. Fan, L. T., Y. W. Huang, and N. Yutani, Chem. Eng. Sci., 41, 189 (1986).
3. Tone, S., H. Seko, H. Maruyama, and T. Otake, J. Chem. Eng. Japan, 7, 44 (1974).
4. Otake, T., S. Tone, M., Kawashima, and T. Shibata, J. Chem. Eng. Japan, 8, 388 (1975).
5. Takami, K., and S. Furusaki, J. Chem. Eng. Japan, 18, 113 (1985).
6. Rowe, P. N., P. U. Foscolo, A. C. Hoffmann, and J. G. Yates, draft paper for engineering foundation conference, "Fluidization IV," (1983).
7. Hoffmann, A. C., and J. G. Yates, Chem. Eng. Commun., 41, 133 (1987).
8. Jin, Y., Z. Q. Yu, Z. W. Wang, and P. Cai, in "Fluidization V," ed. by K. Ostergaard and A. Sorensen, p. 189, Eng. Found., New York (1986).
9. Cai, P., Y. Jin, Z. Q. Yu, and P. Y. Qi, In "Proceedings of the International Symp. on Multi-Phase Flows," p.221, Hangzhou, China (1987).
10. Jin Y., and Z. Q. Yu, International Chemical Engineering, 22, 269 (1982).

# THE EFFECT OF FLOW CONDITIONERS ON THE TENSILE STRENGTH OF COHESIVE POWDER STRUCTURES

H.O. Kono, C.C. Huang and M. Xi ∎ Chemical Engineering Department, West Virginia University, Morgantown, WV 26506
F.D. Shaffer ∎ United States Department of Energy, Pittsburgh Energy Technology Center, Pittsburgh, PA 15236

The tensile strength of the powder structure formed by a cohesive coal powder subjected to low compressive pressures is investigated. By adding small amounts (0.1 to 5.0 wt%) of two typical "flow-conditioner" powders (polysaccharide and calcium stearate), the tensile strength of the coal powder was significantly reduced. The operative mechanism of flow-conditioner particles is explained by proposing an analogy to crystal defects.

## INTRODUCTION

The cohesive characteristics of fine powders [Type C according to the Geldart Classification ([1])] present a critical problem for powder-handling processes such as fluidization, dense-phase pneumatic transport, and mixing. These processes are widely applied in numerous industries, including the food, pharmaceutical, energy, and chemical industries. Recently, fine cohesive powders have been applied in many new technologies such as the production of silicon-based electronic materials, specialty polymers, and highly processed coals.

When a fine cohesive powder is subjected to compressive pressures it usually forms a flocculated-powder structure ([2-4]). The powder structure must be disintegrated before the powder can flow or become fluidized. The strength of the powder structure determines the magnitude of the forces that must be applied to it to initiate flow. Thus, the initial yielding strength of a powder structure may be taken as a measure of the powder's "flowability."

Because fine cohesive powders form powder structures, they do not behave as a viscous liquid or an elastic solid. For example, the shearing stresses that occur in a slowly deforming powder structure can be considered independent of the rate of shear and dependent on the mean compressive pressure acting on the powder structure. In a viscous liquid, this dependecy is reversed. Rather than a viscous liquid or an elastic solid, a fine powder structure behaves as a plastic. Flow is initiated as the powder structure fractures or undergoes plastic deformation.

A number of methods have been developed to measure the mechanical strength of a powder structure. The classic example is the Jenike shear cell. The Jenike shear cell provides a measure of the strength of a powder structure under the high compressive pressures typical of powder storage hoppers. Jenike ([5]) has developed a methodology to design powder storage hoppers based on shear cell data. Molerus ([6,7]) has developed an elaborate theory to describe interparticle forces based on Jenike shear cell measurements. Although the Jenike shear cell is the most widely applied instrument to measure the strength of powder structures, it is not appropriate to accurately measure the strength of delicate powder structures under low-compressive-pressure conditions (<5 kilopascals)([8]). These delicate powder structures can be developed in powder processes such as fluidization, dense-phase pneumatic transport, and mixing.

In this paper, the yielding tensile strength of a powder structure is considered as a measure of the flowability of a fine powder under low compressive pressures. A tensile strength test apparatus developed by Hartley and Parfitt (9) is used to measure the tensile strength of powder structures under low-compressive-pressure conditions. Using this tensile strength measurement, the effect of the two different types of "flow conditioners" on the yielding tensile strength of cohesive powder is studied.

The flow conditioners were polysaccharide and calcium stearate powders. A cohesive Pittsburgh seam coal powder was used as a model cohesive powder. Different reductions in tensile strength were observed for the two flow conditioners: the effectiveness of the calcium stearate peaks at a concentration of about 0.5 - 1.0 wt%, then sharply decreases; the effectiveness of the polysaccharide peaks and levels off at concentrations above 1 wt%. An explanation of the operative mechanism of the flow conditioners is proposed using an analogy with the crystal defect theory developed in solid-state chemistry (10).

## EXPERIMENTAL

Pittsburgh coal is a naturally cohesive powder with a wide size distribution of angularly shaped particles. The Pittsburgh coal used in this study was pulverized to a size consisting of 70% passing through a U.S. No. 200 mesh screen. The size distributions of the Pittsburgh coal and flow-conditioner powders are shown on Figure 1.

Two flow-conditioner powders, calcium stearate and polysaccharide, were added in an attempt to modify the strength of the powder structures formed by the Pittsburgh coal. The flow-conditioner powders were added to the Pittsburgh coal at concentrations ranging from 0.1 to 5.0 wt%. The flow conditioners were mixed with the Pittsburgh coal powder by dispersing the coal powder in a rotary mixer and adding small amounts of flow conditioner at 5-minute intervals. The powder mixture was then passed through a U.S. No. 60 mesh screen three times. Microscopic observation of the powder mixtures indicated that this mixing technique was sufficient to disperse the flow-conditioner particles throughout the coal powder and to disintegrate any agglomerates of the flow-conditioner particles.

The tensile strength was measured using a split-cell apparatus developed by Hartley and Parfitt (9). Powder structures with a range of voidages and strengths were formed by varying the compressive pressure applied to the powder. Because the strength of the powder structure may be sensitive to humidity, the powder mixtures were kept at 370 K for at least three hours before testing. In addition to the tensile strength tests, a scanning electron microscope-spectroscope was employed to visually inspect each of the powder mixtures. This proved useful to assess the degree of interparticle affinity between the Pittsburgh coal and flow-conditioner particles.

## RESULTS AND DISCUSSION

The results of the tensile strength tests are shown in Figure 2. These results are for a voidage corresponding to that of a loosely packed bed condition ($\varepsilon_{lp} \approx 0.5$). The procedure for finding the tensile strength at a loosely packed bed condition involves extrapolation of tensile strength results for lower voidages and is explained in Ref. (2). The test results indicate that the strength of the Pittsburgh coal powder structure can be drastically reduced by the addition of very small amounts of the calcium stearate and polysaccharide powders. This effect could have practical applications and is also of significant interest from the standpoint of developing a fundamental understanding of powder rheology.

The difference in the behavior of the tensile strength results for the two flow-conditioner powders is of particular interest. When polysaccharide is added, the tensile strength of the coal powder structure drops sharply, then levels off after reach-

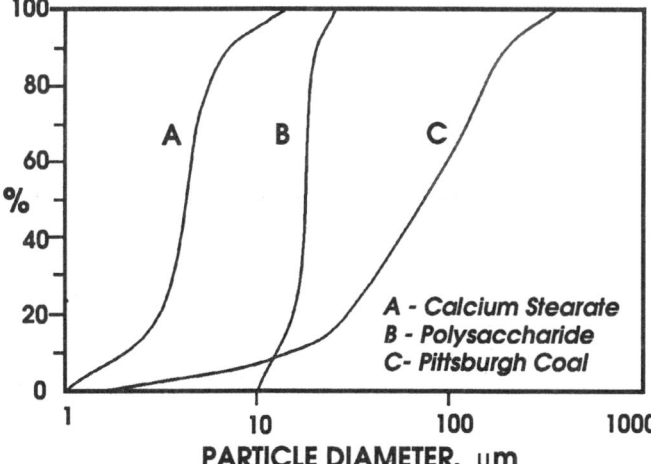

*Figure 1. Size distributions of the Pittsburgh seam coal and flow conditioner powders.*

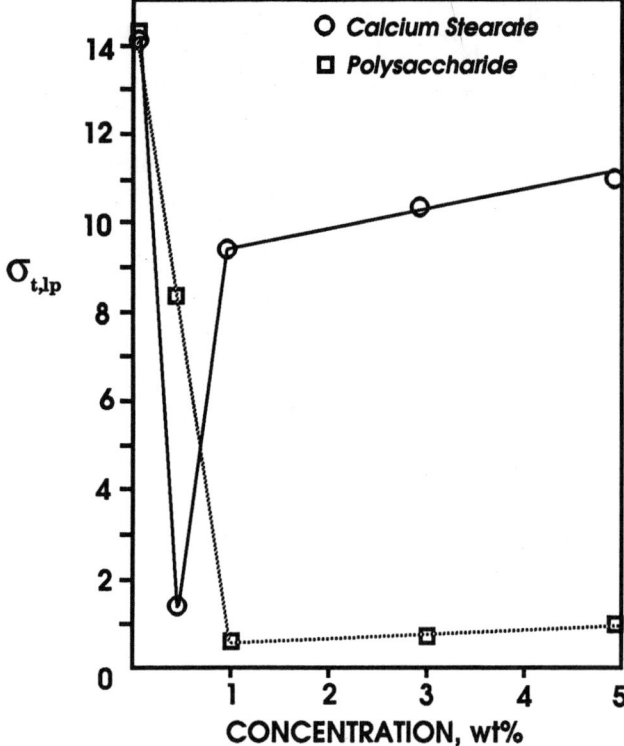

*Figure 2. Tensile strength of loosely packed powder structure ($\sigma_{t,lp}$) as a function of flow-conditioner concentration.*

ing a concentration of 1.0 wt%. The tensile strength also drops sharply when calcium stearate is added up to 0.5 wt% but increases as the concentration increases from 0.5 to 1.0 wt%. The tensile strength then shows a slight increase when the concentration of calcium stearate is further increased to 5 wt%.

Reductions in the strength of food and pharmaceutical powders upon addition of flow-conditioner particles have been observed in previous studies (11,12). The operative mechanism of the flow-conditioner particles was related to their ability to adhere to and coat the surface of the host particles. The flow-conditioner particles then reduced the interparticle forces between the host particles by [1] acting as a physical barrier between the host particles and thus reducing contact forces between the host particles, [2] acting as lubricants to reduce the friction between host particles, or [3] acting as neutralizers of electrostatic charge (12). These previous studies with food and pharmaceutical powders have also suggested that when a powder's strength is reduced via addition of flow conditioners, the bulk density of the powder is also reduced.

The experimental results of this work with a coal powder differ from the results of the above-mentioned studies with food and pharmaceutical powders in that [1] little change in bulk density was associated with the reductions in tensile strength of the coal powder (see Figure 3), and [2] polysaccharide particles do not adhere to the surface of coal particles. Also, in previous studies, no explanation is given as to how flow-conditioner particles reduce the net strength of a powder when the concentration of flow conditioner is often so low (e.g., 0.1 wt%) that there are only enough flow conditioner particles to interact with a small fraction of the host particles.

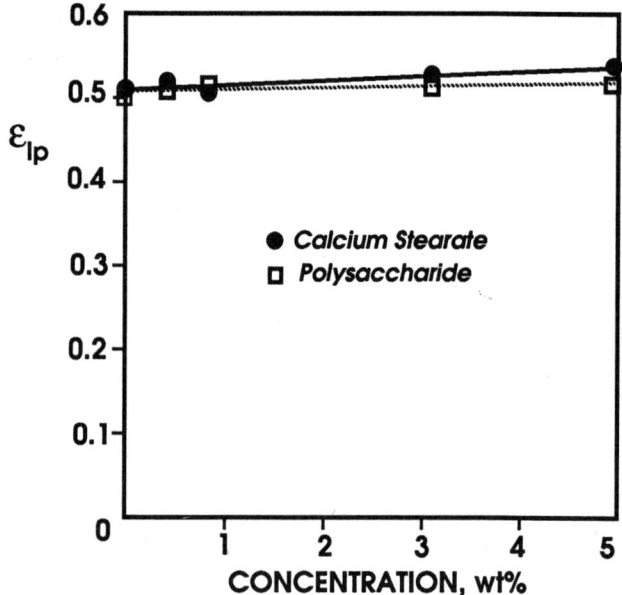

*Figure 3. Voidage of loosely packed powder mixtures ($\varepsilon_{lp}$) as a function of flow-conditioner concentration.*

In this paper, to elucidate the operative mechanism by which flow-conditioners are effective at such low concentrations, we suggest that the flow-conditioner particles reduce the net tensile strength of a powder structure by creating scattered weak points throughout the powder structure. The weak points serve to initiate fractures throughout the powder structure. As soon as enough fractures are initiated to disintegrate the powder structure, flow or fluidization may be initiated. Thus, it is only necessary for the flow-conditioner particles to interact with a fraction of the host particles.

An analogy can be drawn between this explanation and the explanation of the role of defects to modify the stength of crystalline structures. The explanation of crystal defects states that defects can be created in a crystalline structure by adding small amounts of an impurity (10). The defects then reduce the net strength of the crystalline structure by creating weak points that initiate fractures. Even though a powder structure is very different from a crystalline structure, e.g., the particles are much larger than a molecular size and do not form an ordered structure, the flow conditioners are analgous to the impurities in that they also create weak points that initiate fractures, and they can also be effective at very low concentrations (< 1 wt%).

The ability of polysaccharide particles to create weak points may be attributed to their low surface affinity with coal particles. Scanning electron microscope observations reveal that the polysaccharide particles have a low affinity amongst themselves and with the coal particles; and therefore, they do not adhere to the coal particles (see Photograph 1). Thus, the polysaccharide particles create points with tensile strengths much lower than the tensile strength of the surrounding coal particles. A concentration of less than 1 wt% of polysaccharide is required to initiate enough fractures in the coal powder structure to achieve a maximum reduction in tensile strength. Further additions of polysaccharide have little effect on the tensile strength (see Figure 2).

The calcium stearate particles form weak points in the coal powder structure but by a different mechanism than the polysaccharide. Scanning electron microscope observations reveal that calcium stearate particles have a high affinity amongst themselves and the coal particles. This strong interparticle affinity gives pure calcium stearate a high tensile strength. Photograph 2 shows that smaller calcium stearate particles do adhere to the surface of a larger coal particle. Spectroscopic analysis was used to verify that most of the par-

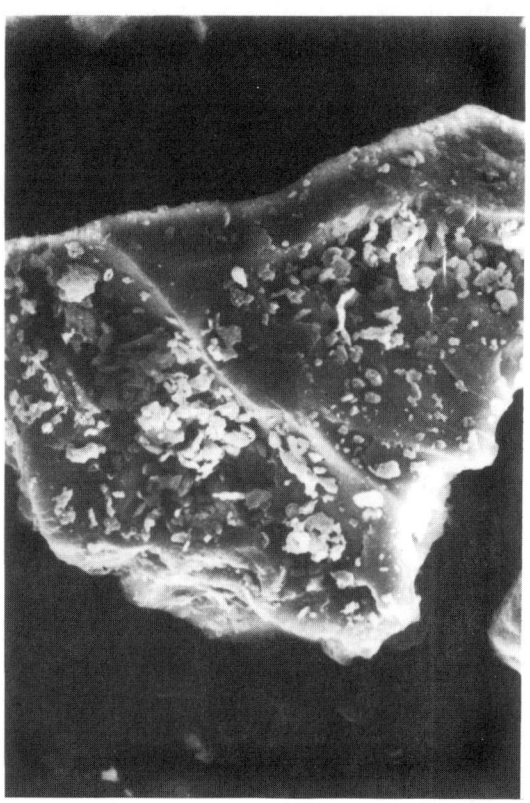

*Photograph 1. Scanning Electron Microscope picture showing a mixture of polysaccharide and coal particles.*

*Photograph 2. Scanning Electron Microscope picture showing calcium stearate particles adhering to the surface of a coal particle.*

ticles shown coating the surface of the coal particle in Photograph 2 were indeed calcium stearate.

We suggest that the calcium stearate particles create weak points by lubricating the surface of coal particles. The lubricating ability of calcium stearate has also been reported in other studies (11,12). As small amounts, up to 0.5 wt%, of calcium stearate were added, most of the calcium stearate particles adhered to the surfaces of coal particles. As the concentration was increased to 0.5 wt%, the coal powder became "saturated" with calcium stearate particles. As the concentration was increased above 0.5 wt%, the additional calcium stearate particles were precipitated (i.e., did not adhere to the coal particles) and tended to form agglomerates amongst themselves. Because of the high tensile strength of pure calcium stearate, these agglomerates simply added to the net tensile strength of the powder structure. Thus, the minimal point of the tensile strength curve for the calcium stearate is interpreted as corresponding to the point of saturation of calcium stearate particles in the coal powder; and the subsequent rise in tensile strength is interpreted as corresponding to the precipitation of calcium stearate particles into agglomerates.

## DISCLAIMER

Reference in this report to any specific commercial product, process or service is to facilitate understanding and does not necessarily imply its endorsement or favoring by the United States Department of Energy.

## LITERATURE CITED

1. Geldart, D., "Types of Gas Fluidization," Powder Tech., 285 (1973).
2. Kono, H.O., Chiba, S., Ells, T., and Suzuki, M., "The Effect of Emulsion Phase Properties on Fluidization," Powder Tech. 48, 51-58 (1986).
3. Kono, H.O., Ells, T., Chiba, M., Suzuki, M., and Morimoto, E., "Quantitative Criteria for Emulsion Phase Characteristics and for the Transition between Particulate and Bubbling Fluidization," Powder Tech. 52, 69-76 (1987).
4. Kono, H.O., Huang, C.C., Xi, M., Nakayama, T., and Hikosaka, T., "Characterization of Fine Powders by Rheological Properties of the Gas-Powder Colloid Structure at Aerated Conditions," AIChE Symp. Ser., No. 262, Vol. 84, 74-81 (1988).
5. Jenike, A.W., "Storage and Flow of Solids," Bulletin of the Univ. of Utah, Vol. 53, No. 26 (1964).
6. Molerus, O., "Theory of Yield of Cohesive Powders," Powder Tech., 12, 259-275 (1975).
7. Molerus, O., "Effect of Interparticle Cohesive Forces on the Flow Behavior of Powders," Powder Tech., 20, 161-175 (1978).
8. Eckhoff, R.K., Leverson, P.G., and Schubert, H., "The Combination of Tensile Strength Data of Powders and Failure Loci from Jenike Shear Cell Tests," Powder Tech., 19, 115-118 (1978).
9. Hartley, P.A., and Parfitt, G.D., "An Improved Split-Cell Apparatus for the Measurement of Tensile Strength of Powders," J. Phys. E. Sci. Instr., 17, 347 (1984).
10. West, A.R., Basic Solid State Chemistry, p. 206, John Wiley & Sons, Chitester, NY (1984)
11. Peleg, M., and Mannheim, C.H., "Effect of Conditioners on the Flow Properties of Powdered Sucrose," Powder Tech., 7, 45-50 (1973).
12. Hollenbach, A.M., Peleg, M., and Rufner, R., "Interparticle Surface Affinity and the Bulk Properties of Conditioned Powders," Powder Tech., 35, 51-62 (1983).

# TRANSIENTS IN BED EXPANSION OF A THREE-PHASE FLUIDIZED BED

Y-M Chen, F. Bavarian and L.-S. Fan ■ Department of Chemical Engineering, The Ohio State University, Columbus, OH 43210
R.D. Buttke and W.I. Beaton ■ Amoco Oil Company, Naperville, IL 60566

Experimental study of the transient behavior of a gas-liquid-solid fluidized bed operated in the dispersed bubble flow regime is conducted. The transient experiment is initiated by introducing a step increase or decrease in the liquid velocity while maintaining the gas velocity constant. For a three-phase fluidized bed operated at low gas flow rates, the dynamic responses to both a step increase and decrease in the liquid velocity are very similar to those of a liquid-solid fluidized bed. With a step decrease, the bed level is shown to decrease linearly in the entire transient period and the system quickly reaches the final steady state. With a step increase, the bed level is shown to increase with an initial linear period followed by a long tail-off period and the system reaches the final steady state much slower. However, for a three-phase fluidized bed operated at moderate and high gas flow rates, bed responses to a step increase and decrease in the liquid velocity both show a tail-off period and similar response times are observed. A simple theoretical model is developed to describe the transient behavior of bed response in the gas-liquid-solid fluidized bed.

Transients in a gas-liquid-solid fluidized bed may occur under a number of situations. During start-up and shut-down of a system, transients are expected; even during normal operation, fluctuations in the feeding stream are inevitable which also cause transients to occur. Many operational problems, such as particle entrainment and bed overshoot, are directly associated with the hydrodynamics of the transients. Others, such as local overheating, can be the consequences of transients.

The study of transient behavior of fluidized beds has been confined to liquid-solid and gas-solid systems in the literature. In Slis et al.'s ([1]) study with a liquid-solid fluidized bed, it was found that fundamental differences exist between the transients responding to a step increase and a step decrease in the liquid velocity. With a step decrease, the boundary of the holdup discontinuity was shown to remain sharp whereas the boundary broadened with a step increase. It was also found that time required to reach a final steady state is much shorter for a step decrease than that for a comparable step increase. These differences were attributed to the formation of a shock wave. The continuity wave velocity of a liquid-solid medium was shown to increase as the liquid holdup decreases. When a step decrease in the liquid velocity was introduced, the system consisted of a region of high liquid holdup (slow wave) above a region of low liquid holdup (fast wave). The fast wave overtakes the slow one and forms a shock wave, rendering the holdup discontinuity sharp ([2]). Fan et al. ([3]), using a linearized model, extended the dynamic study of a liquid-solid fluidized bed responding to a variety of liquid velocity changes, including step, pulse and sinewave. Didwania and Homsy ([4]) employed the transient bed expansion of a liquid-solid fluidized bed to study Rayleigh-Taylor instability; however, the instability was not observed except for the extreme cases that the upper portion of the bed was not fluidized. Little is known so far about the transient of a gas-liquid-solid fluidized bed.

In the present study, the transient experiments of a gas-liquid-solid fluidized bed are conducted by introducing a step increase or decrease in the liquid

---

*To whom correspondence should be addressed.

velocity while maintaining the gas velocity constant. The system is operated in the dispersed bubble flow regime, with small bubbles and high gas holdups, by using a 0.3 wt% t-pentanol aqueous solution as the liquid phase. This particular condition is selected in order to simulate the industrial operation of hydrotreating units operated under high pressure and temperature conditions. It was found by Fan et al. (5) that an annular fluidized bed operated under an ambient condition and used a 0.5 wt% t-pentanol aqueous solution as the liquid phase could successfully simulate the EDS Coal Liquefaction Process conditions reported by Tarmy et al. (6). A simple theory is also developed to account for the transient behavior of bed response to the step change.

## FORMULATION

Consider a three-phase fluidized bed operated in the dispersed bubble flow regime with initial gas and liquid superficial velocities of $U_{g0}$ and $U_{\ell 0}$, and phase holdups of $\epsilon_{g0}$ and $\epsilon_{\ell 0}$. At a certain instance, step changes in the superficial velocities, $U_{g1}$ and $U_{\ell 1}$, are introduced. After a certain period of time, the system eventually reaches a final steady state with new phase holdups of $\epsilon_{g1}$ and $\epsilon_{\ell 1}$ throughout the bed. During the transient, the phase holdups ($\epsilon_g$, $\epsilon_\ell$, $\epsilon_s$) and the linear velocities of the individual phases ($v_g$, $v_\ell$, $v_s$) are functions of both time and position. The overall continuity requires that the summation of volumetric flow rates of individual phases be equal to the overall superficial velocities:

$$\epsilon_g v_g + \epsilon_\ell v_\ell + \epsilon_s v_s = U_{\ell 1} + U_{g1} \quad (1)$$

In Equation (1), the gas phase is assumed to be incompressible. The summation of all the phase holdups necessitates that

$$\epsilon_g + \epsilon_\ell + \epsilon_s = 1 \quad (2)$$

When simulating the dynamic response of a three-phase fluidized bed, a fundamental difficulty is encountered. Unlike the liquid-solid fluidized system that the relative motion between the phases is well described by the Richardson and Zaki (7) equation, such equations governing the relative motions are lacking for a gas-liquid-solid fluidized system. Consequently, only part of the dynamic response of the bed expansion can be formulated at the present stage.

Assume that in a three-phase fluidized bed the relative velocities between phases are dependent only upon phase holdups:

$$v_\ell - v_s = f(\epsilon_g, \epsilon_\ell) \quad (3)$$

$$v_g - v_\ell = k(\epsilon_g, \epsilon_\ell) \quad (4)$$

where f and k are two unknown functions of phase holdups $\epsilon_g$ and $\epsilon_\ell$. The forms adopted in Equations (3) and (4) are consistent with those of gas-liquid, gas-solid and liquid-solid systems found in the literature. It is also assumed that the equilibrium conditions of Equations (3) and (4) are applicable to the three-phase fluidized system under dynamic condition. A similar equilibrium assumption was made and verified by Slis et al. (1) for a dynamic liquid-solid fluidized system. By combining Equations (1) through (4) to eliminate $\epsilon_s$, $v_g$ and $v_\ell$, the linear velocity of the solid phase can be expressed as

$$v_s = U_{\ell 1} + U_{g1} - \epsilon_\ell f(\epsilon_g, \epsilon_\ell) - \epsilon_g [f(\epsilon_g, \epsilon_\ell) + k(\epsilon_g, \epsilon_\ell)] \quad (5)$$

Since the boundary between the fluidized bed and the freeboard regime is marked by the particles, the rate of change of bed height is equal to the particle linear velocity at the top of the bed. From Equation (5), one has

$$\frac{dh}{dt} = v_s^* = U_{\ell 1} + U_{g1} - \epsilon_\ell^* f(\epsilon_g^*, \epsilon_\ell^*)$$

$$- \epsilon_g^* [f(\epsilon_g^*, \epsilon_\ell^*) + k(\epsilon_g^*, \epsilon_\ell^*)] \quad (6)$$

where h is the bed height and the superscript "*" indicates the local value at the top of the bed.

Consider a steady state operation of <u>another</u> three-phase fluidized bed with superficial velocities of $U_g^*$ and $U_\ell^*$ such that this system has holdups of $\epsilon_g^*$ and $\epsilon_\ell^*$ throughout the bed. Since the linear velocity of the solid phase is zero at steady state operation, the equilibrium conditions in this system require that

$$\frac{U_\ell^*}{\epsilon_\ell^*} = v_\ell^* = f(\epsilon_g^*, \epsilon_\ell^*) \quad (7)$$

$$\frac{U_g^*}{\epsilon_g^*} - \frac{U_\ell^*}{\epsilon_\ell^*} = v_g^* - v_\ell^* = k(\epsilon_g^*, \epsilon_\ell^*) \quad (8)$$

Combining Equations (6), (7) and (8), one has

$$\frac{dh}{dt} = v_s^* = U_{\ell 1} + U_{g1} - (U_g^* + U_\ell^*) \quad (9)$$

Equation (9) indicates that the velocity of the bed height response is equal to the difference between the sum of the step-changed superficial velocities and the sum of the superficial velocities which produce the same holdup conditions at steady state operation as those at the top of the dynamic bed.

After the step changes are made, the holdup conditions at the top of the bed will remain as $\epsilon_{g0}$ and $\epsilon_{\ell 0}$ for a period of time until the first continuity wave reaches the top. During this period, the velocity of the bed height response is given by replacing $(U_g^* + U_\ell^*)$ by $(U_{g0} + U_{\ell 0})$ in Equation (9):

$$\frac{dh}{dt} = U_{\ell 1} + U_{g1} - (U_{g0} + U_{\ell 0}) \quad (10)$$

When the final steady state is established, the holdup condition at the top of the bed becomes $\epsilon_{g1}$ and $\epsilon_{\ell 1}$, and Equation (9) requires that the bed height is a constant because

$$\frac{dh}{dt} = U_{\ell 1} + U_{g1} - (U_{g1} + U_{\ell 1}) = 0 \quad (11)$$

which is consistent with the steady state condition. Under the extreme condition of zero gas flow rate, Equation (9) reduces to the equation given by Slis et al. (<u>1</u>) and Didwania and Homsy (<u>4</u>) for a liquid-solid fluidized bed.

## EXPERIMENTAL

A schematic diagram of the apparatus for transient experiments of a three-phase fluidized bed is shown in Figure 1. In the experiments, the fluidized bed was operated in a acrylic column 122 cm in height and 15.2 cm in diameter. Ten pressure taps along the column were connected to manometers to measure the pressure distribution. The distributor consisted of three sections: the plenum chamber, the gas-liquid distributor, and a fixed bed section. The plenum chamber was filled with 1 cm polyethylene spheres. Liquid first entered the plenum chamber and then flowed into the gas-liquid distributor through forty tubes (6 mm ID) which ended in 1.6 mm injection holes into the fixed bed section. Gas entered the distributor through two ports located on the side wall and passed through 85 injection holes, 1 mm in diameter, into the fixed bed section.

Two solenoid valves were installed along the liquid stream line such that part of the liquid flow could be recycled back to the surge tank upon opening either solenoid valve. The experiments were performed by setting the liquid and gas flow rates at desired values and then by turning the solenoid valve(s) off or on, a step increase or decrease in the liquid flow rate was made, respectively. During the change of liquid flow, the gas flow rate was maintained constant. The dynamic change in the bed height responding to the step change in liquid flow was recorded by a video camera.

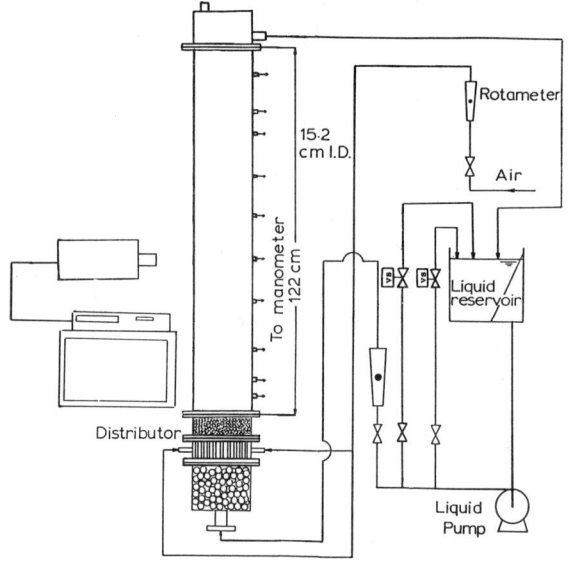

Figure 1  A schematic diagram of the apparatus for transient experiments of a three-phase fluidized bed.

In order to operate the system in the dispersed bubble flow regime, with small bubbles and high gas holdups, a 0.3 wt% t-pentanol aqueous solution was used as the liquid phase. The choice of the liquid was based on the studies of Fan et al. (5) and Gorowara (8). In the present experiments, air and 1.5 mm glass beads were used as the gas and solid phases, respectively. Approximately 5% of the glass beads were painted black to improve visualization. Details of the experimental conditions are given in Table 1.

## RESULTS AND DISCUSSION

Figure 2 shows a typical result of bed height response to a step decrease in the liquid flow rate while maintaining a low, constant gas flow rate (Run 1a). As shown in the figure, the bed collapses immediately after the introduction of the step decrease. The descending velocity of the bed level is found very close to a constant, and 5.4 seconds after the step change, the final steady state is established. The entire process of

TABLE 1.  Summary of operation conditions of the liquid and gas flow rates and the steady state phase holdups.

| Run | $U_g$(cm/s) | $U_{\ell 0}$(cm/s) | $U_{\ell 1}$(cm/s) | $\epsilon_{g0}$(-) | $\epsilon_0$(-) | $\epsilon_{g1}$(-) | $\epsilon_1$(-) |
|---|---|---|---|---|---|---|---|
| 1 a | 1.01 | 7.03  | 3.00  | 0.081 | 0.712 | 0.130 | 0.513 |
| 1 b | 1.01 | 3.00  | 7.03  | 0.130 | 0.513 | 0.081 | 0.712 |
| 2 a | 1.01 | 8.75  | 3.15  | 0.050 | 0.747 | 0.144 | 0.550 |
| 2 b | 1.01 | 3.51  | 8.75  | 0.144 | 0.550 | 0.050 | 0.747 |
| 3 a | 1.01 | 10.5  | 4.10  | 0.045 | 0.781 | 0.123 | 0.610 |
| 3 b | 1.01 | 4.10  | 10.5  | 0.123 | 0.610 | 0.045 | 0.781 |
| 4 a | 2.00 | 7.03  | 3.00  | 0.137 | 0.730 | 0.127 | 0.559 |
| 4 b | 2.00 | 3.00  | 7.03  | 0.127 | 0.559 | 0.137 | 0.730 |
| 5 a | 2.00 | 8.76  | 3.50  | 0.128 | 0.756 | 0.188 | 0.629 |
| 5 b | 2.00 | 3.50  | 8.76  | 0.188 | 0.629 | 0.128 | 0.756 |
| 6 a | 2.00 | 10.50 | 4.20  | 0.101 | 0.785 | 0.189 | 0.652 |
| 6 b | 2.00 | 4.20  | 10.50 | 0.189 | 0.652 | 0.101 | 0.785 |
| 7 a | 4.00 | 7.03  | 3.08  | 0.253 | 0.766 | 0.333 | 0.698 |
| 7 b | 4.00 | 3.08  | 7.03  | 0.333 | 0.698 | 0.253 | 0.766 |
| 8 a | 4.00 | 8.75  | 3.76  | 0.218 | 0.811 | 0.354 | 0.723 |
| 8 b | 4.00 | 3.76  | 8.75  | 0.354 | 0.723 | 0.218 | 0.811 |

the bed response shown in Figure 1 can be well represented by a linear decrease of the bed height from its initial level (51 cm) to the final level (30 cm).

Figure 3 shows a corresponding result of bed height response to a step increase in the liquid velocity while maintaining the gas velocity constant (Run 1b). Note that the initial operation condition of Figure 2 is chosen to be the same as the final operation condition of Figure 3, and vice versa. As shown in Figure 3, the bed response to a step increase in the liquid velocity is significantly different from that of Figure 2. The bed height does not respond immediately to the step increase, and shows a delay of about 1.3 seconds before expansion. This time delay is due mainly to the response of the solenoid valve. When expanding, the bed level shows a initial period of linear increase followed by a long tail-off period. Consequently, the time required to reach the final steady state is much longer than that of Figure 2. From Figures 2 and 3, it is found that the characteristics of bed response for a three-phase fluidized bed at low gas velocities are very similar to those observed by Slis et al. (1) for the liquid-solid fluidized bed. This result is not very surprising because at low gas flow rates the effects of the gas flow on the liquid-solid medium is relatively insignificant.

Figure 4 (Run 4a and 4b) and Figure 5 (Run 8a and 8b) show two dynamic responses to a step decrease and increase in the liquid velocity at moderate and high gas flow rates, respectively. It is noted that the bed responses to a step increase are similar to Figure 3, but the bed responses to a step decrease are different from Figure 2. As shown in Figures 4 and 5, the descending velocity of the bed height responding to a step decrease no longer remains constant over the entire process but exhibits a tail-off section as the final steady state is approached. The phenomenon of tail-off is more pronounced as the gas velocity becomes higher, as shown in Figure 5. Consequently, the total response times for a step decrease and increase become comparable. This may be due to the fact that when a step decrease in the liquid velocity is introduced, the gas holdup in the lower portion of the bed becomes higher. The increase in the gas holdup due to

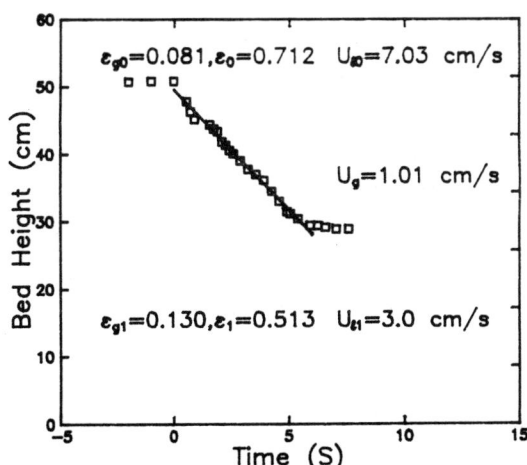

Figure 2 The bed height response to a step decrease in liquid velocity at a low gas flow rate (Run 1a).

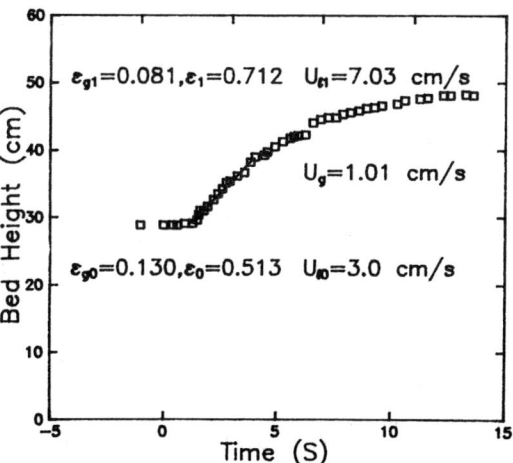

Figure 3 The bed height response to a step increase in liquid velocity at a low gas flow rate (Run 1b).

the decrease in the liquid flow is more pronounced at higher gas flow rates. The lower portion of the bed with a lower liquid holdup but a higher gas holdup does not guarantee that the wave propagation velocity in that portion is higher. Hence, a shock wave, which would result in the linear descending bed level for the entire period, may not necessarily exist when a step decrease in the liquid velocity is introduced into a three-phase fluidized bed. This result indicates that the transient behavior of a three-phase fluidized bed is in general different from that of a liquid-solid fluidized system.

In Figure 6, predictions of Equation (10) are compared with experimental measurements of both the linear ascending and descending velocities of bed height, due respectively to step increase and decrease in the liquid velocity. In the case of the dynamic response to a step increase in the liquid velocity, the theory only covers the initial linear section, which is about half of the entire response. In the case of the dynamic response to a step decrease in the liquid velocity at low gas flow rates, such as Figure 2, the theory essentially covers the entire response. For the case of the dynamic response to a step decrease in liquid velocity at moderate and high gas flow rates, Figures 4 and 5, the theory only covers the initial linear section, which consists of the major portion of the response. The reason for not including the tail-off period is due to the fact that in that period the

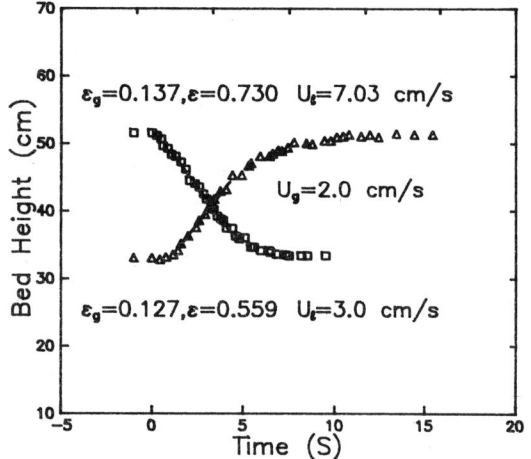

Figure 4 Bed height responses to step decrease (Run 4a) and increase (Run 4b) in the liquid velocity at a moderate gas flow rate.

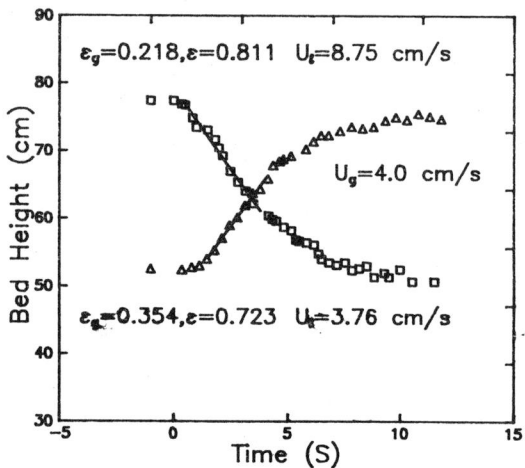

Figure 5 Bed height responses to step decrease (Run 8a) and increase (Run 8b) in the liquid velocity at a high gas flow rate.

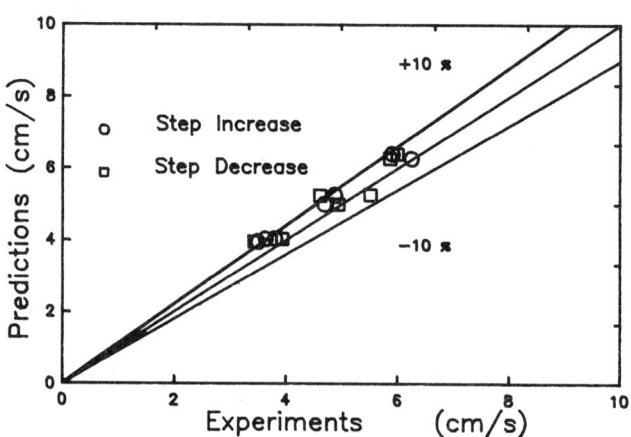

Figure 6 Comparison of predictions by Equation (10) and measurements of both linear ascending and descending velocities of bed level

holdup conditions at the top of the bed are not expected to remain as $\epsilon_{g0}$ and $\epsilon_{\ell 0}$, and Equation (10) can no longer be applied. As shown in Figure 6, the predictions are in good agreement with all the measurements. The discrepancies are within 10 % in most cases.

CONCLUSION

For a three-phase fluidized bed operated at low gas low rates, the dynamic responses to both a step increase and decrease in the liquid velocity are very similar to those of a liquid-solid fluidized bed. With a step decrease, the bed level is shown to decrease linearly in the entire transient and the system quickly reaches the final steady state. With a step increase, the bed level is shown to increase with an initial linear period followed by a long tail-off period and the system reaches the final steady state much slower. However, for a three-phase fluidized bed operated at moderate and high gas flow rates, bed responses to a step increase and decrease in the liquid velocity both show a tail-off period and response times for the two transients become comparable. The proposed theory predicts the linear period, which covers the major portion of the response to step decrease and part of the response to step increase, to within 10 % error.

NOTATION

Roman Letters

| | |
|---|---|
| f | unknown function for the relative motion between the liquid and solid phases, $L/\theta$ |
| h | height of the three-phase fluidized bed, L |
| k | unknown function for the relative motion between the gas and liquid phases, $L/\theta$ |
| t | time, $\theta$ |
| U | superficial velocity, $L/\theta$ |
| v | linear velocity of individual phase, $L/\theta$ |

Greek Letters

| | |
|---|---|
| $\epsilon$ | bed voidage, - |
| $\epsilon_g$ | gas holdup, - |
| $\epsilon_\ell$ | liquid holdup, - |
| $\epsilon_s$ | solid holdup, - |
| $\rho$ | density, $M/L^3$ |

Subscripts

| | |
|---|---|
| g | gas phase |
| $\ell$ | liquid phase |
| s | solid phase |
| 0 | initial value |
| 1 | final value |

Superscript

| | |
|---|---|
| * | local value at the top of the bed |

LITERATURE CITED

1. Slis, P. L., T. W. Willemse, and H. Kramers, "The response of the level of a liquid fluidized bed to a sudden change in the fluidizing velocity," Appl. Sci. Res., 8, 209 (1959).

2. Wallis, G. B., One-Dimensional Two-Phase Flow, McGrow Hill Inc., New York, 122 (1969).

3. Fan, L. T., J. A. Schmitz, and E. N. Miller, "Dynamic of liquid-solid fluidized bed expansion," AIChE J., 9, 149 (1963).

4. Didwania, A. K. and G. M. Homsy, "Rayleigh-Taylor instability in fluidized bed," Ind. Eng. Chem. Fundam., 20, 318 (1981).

5. Fan, L.-S., F. Bavarian, R. Gorowara, B. E. Kreischer, R. D. Buttke, and L. E. Peck, "Hydrodynamics of gas-liquid-solid fluidization under high gas holdup conditions," Powder Tech., 53, 285 (1987).

6. Tarmy, B. L., M. Chang, C. A. Coulaloglou, and P. R. Ponz, "The three phase characteristics of the EDS coal liquefaction reactor: their development and use in reactor scaleup," Proc. 8th Int. Symp. Chem. Reaction Eng., 303 (1984).

7. Richardson, J. F. and W. M. Zaki, "Sedimentation and fluidization: part I," Trans. Inst. Chem. Engr., 32, 35 (1954).

8. Gorowara, R., "Effects of surfactants on three-phase fluidized bed hydrodynamics," M.S. Thesis, The Ohio State University, Columbus, Ohio (1988).

# AN OVERVIEW OF MATERIAL SYNTHESIS BY AEROSOL PROCESSES

Sotiris E. Pratsinis ■ Department of Chemical and Nuclear Engineering, Center for Aerosol Processes, University of Cincinnati, Cincinnati, OH 45221-0171

Aerosol processes are attractive for the manufacture of large volumes of particulates with prescribed physicochemical characteristics. This paper reviews the current state of aerosol processes in manufacture of particulate commodities (carbon blacks and inorganic pigments) and high technology materials (optical waveguides, thin films for microelectronics and advanced ceramics). The lack of nonintrusive characterization instruments for highly concentrated aerosols in nonuniform, nonisothermal environments and a paucity of correlations between process variables and particulate characteristics place limitations to the development, design, scale up and control of industrial aerosol processes. Nevertheless, though much remains to be learned about the fundamentals of aerosol formation and growth, enough is known to apply basic engineering principles to the design of industrial aerosol processes.

## INTRODUCTION

Aerosol is every suspension of solid particles and/or liquid droplets in gases ([1]). Individual particulate sizes may range from molecular clusters (<10A) up to 10 μm particles ([2]). An aerosol process is any operation or series of operations that causes a physical or chemical change to a gaseous suspension of solid or liquid particulates. Traditionally aerosol science has been devoted to the prevention of the deleterious effects of aerosols on human health and environment. Aerosol engineering is a new discipline aiming to develop industrial aerosol processes for manufacture of advanced materials and particulate commodities.

Aerosol processes have been used in industry to make carbon blacks, pigments, polymer additives and reinforcing agents ([3]). These processes have also found application in removal of pollutant gases or particles from process streams by scrubbing, spraying, filtration, electrostatic precipitation, sedimentation and cyclone separation ([4]). More recently, aerosol processes find use in fabrication of materials for high technology applications such as optical waveguides ([5,6]), thin films for microelectronics ([7]) and ceramic powders for advanced ceramics ([8]). These processes, however, are not always desirable since they are responsible for particulate air pollution ([9]), corrosion/erosion of industrial equipment ([10]) and reduction of industrial process yields ([11,12]).

Aerosol processes are advantageous for manufacture of particulates since they do not require the several tedious processes and high liquid volumes of wet chemistry. Aerosol processes can be used when high purity is of importance such as in fabrication of optical waveguides. Aerosol technology ('vapor' deposition processes) replaced the double crucible technology for manufacture of low loss fibers for telecommunications ([13]). Aerosol processes do not require surfactants to make fine, spherical particles which are important in manufacture of superalloys and highly tough ceramics. In aerosol processes effective reactant mixing is achieved in much shorter time scales than in wet chemistry processes. Finally, aerosol processes are energy efficient operations for manufacture of particulates. A typical example is the production of silicon for photovoltaic applications in an aerosol reactor where the power consumption per unit mass of product silicon can be an order of magnitude lower than in the conventional Siemens process (14).

In every aerosol process, several physicochemical phenomena may simultaneously take place: transport of heat and mass, chemical reactions, particle nucleation, condensation, coagulation, thermophoresis and diffusion. Depending on aerosol composition and individual particle size, various mechanisms can be important for a specific physicochemical phenomenon such as particle nucleation according to classical Becker-Doering or Lothe-Pound theories (15), chemical kinetics far or close to equilibrium and particle growth in the free molecular or continuum regime. Since many physicochemical phenomena and mechanisms may be involved in any aerosol process, a large number of variables affects the product aerosol characteristics.

Frequently, aerosol scientists and engineers focus on optimization of a specific characteristic of the product aerosol such as deposition efficiency in lightguide fabrication, size uniformity in powder manufacture for advanced ceramics or opacity per unit mass in production of pigments. In all these applications the design and operation of the specific industrial aerosol processes currently rely on experience and empiricism. In flame aerosol reactor technology, for example, though several patents on reactor design, feed additives and reactant mixing have been filed, the fundamentals of these systems have not been understood (16). As a result, it comes as no surprise that a recent engineering survey found that the particulate industry in U.S. has made practically no progress in the last 20 years (17).

On the other hand, most basic studies of aerosol processes have focussed on the investigation of a specific physicochemical system without aiming toward the development of universal relationships between process variables and product characteristics in terms of dimensionless numbers and nomographs. Notable exceptions are the analysis of stirred tank aerosol reactors (18,19), the analysis of aerosol formation by condensation in laminar flows (20,21), a Levenspiellian analysis of standard chemical reactors with respect to product aerosol characteristics: concentration, average size, polydispersity, surface area and yield (22), and the analysis of the effect of reactant gas mixing on product aerosol characteristics in tubular flow reactors (23). Although these studies were developed for rather idealistic laboratory scale systems, they provided valuable insight for the design of industrial aerosol processes (24).

This paper reviews the current state of aerosol processes in manufacture of commodity and high technology materials. Emphasis is placed on fundamental experimental and theoretical studies as they are related to manufacture of carbon blacks, ceramic powders, optical fibers and thin films for microelectronics.

## CARBON BLACKS

Carbon blacks are the oldest manufactured aerosols. Carbon blacks are made primarily by the oil furnace process in which an aromatic residual oil ("feedstock") is sprayed into a hot gas-air flame in a furnace where carbon black is formed as soot. The flue gas is rapidly quenched and the soot is collected by electrostatic precipitators, cyclones and bag house filters (25). The product carbon blacks appear as aggregates which contain several spherical particles. The employed stoichiometric fuel:oxidant ratio is usually about 2:1 (26).

Depending on the commercial application, various grades of carbon blacks are made by empirically controlling the elementary particle size and degree of aggregation. Production of fine particles requires short contact times and high temperatures while coarse particles are made by long contact times and lower temperatures (26). The turbulence and the residence time in the furnace are controlled empirically by the air flow rate and the length-to-diameter ratio of the furnace. The extent of particle aggregation is reduced by addition of traces of potassium salts to the flame (27).

Although substantial progress has been made in understanding the fundamentals of fuel lean flames, the mechanism of soot formation in fuel rich flames such as those encountered in manufacture of carbon blacks has not been elucidated (28). The environmental implications (toxicity and visibility reduction) of released soot and the control of particle and aggregate size of carbon blacks motivate the research of soot formation and growth. As a result several basic and applied studies have been carried out to elucidate the important physicochemical phenomena or the process parameters controlling soot dynamics. Formation of soot has been studied extensively in premixed flames (29,30), stagnation point diffusion flames (31), laminar diffusion flames (32) and in shock tubes (33).

Studies of soot formation in laminar diffusion flames are especially important since these flames retain some of the key features (e.g. reactant mixing) of industrial processes for manufacture of carbon blacks. It has been established that during hydrocarbon oxidation in laminar diffusion flames, soot particles are formed (incepted) by cyclization reactions at the fuel rich side of the flame. The soot particles grow by coagulation and surface reactions (material addition from the gas phase) and the total soot particle concentration decreases as the mass increases. At the end of the flame, the soot is depleted by oxidation but the flame temperature is rapidly reduced (quenched) and, thus, soot particles are produced. At the later part of the flame the total soot mass decreases while the total particle concentration may either remain intact when oxidation takes place at the particle surface (e.g. by OH radicals) or increase when $O_2$ diffuses into and disintegrates the particle (34).

Soot particles are characterized by in situ laser scattering/extinction, dissymmetry (35), fluorescence (34), thermophoretic (36) and microscopic analyses (29,37). Laser measurements of soot concentration and size are popular since they are nonintrusive and can be rapidly made. They have, however, two major drawbacks: (a) an assumption must be made for the shape of the soot size distribution to facilitate inversion of the scattering/extinction measurements and recover the true soot size distribution and (b) the refractive index of the soot must be known at each point of the measurement.

The first drawback is usually overcome by assuming a monodisperse, self-preserving or lognormal shape of the soot size distribution. These assumptions may, however, result in erroneous estimates of the true soot particle concentration and average size (38). The strong dependence of scattering on the sixth power of the particle diameter makes the laser measurements very sensitive to the tail of the size distribution. These measurements are especially uncertain at the early stages of soot formation where the size distribution is not well represented by lognormal or self-preserving formulations. Nevertheless, simulations of soot aerosol dynamics have shown that the choice of the initial soot size distribution does not affect the predicted soot particle characteristics downstream of the flame front as long as the assumed initial soot size distribution is consistent with optical data (30).

The second drawback is usually overcome by utilizing an experimentally measured refractive index for hydrocarbon (for example propane or acetylene) or arc lamp carbon soots throughout the flame. This procedure does not result in an error more than 20% on the estimated average particle diameter (35). Laser techniques have been successfully used to measure particle concentrations as high as $10^{13}$ particles/cm$^3$ and sizes down to 50 A.

Understanding the evolution of the soot size distribution as a function of time is important for proper inversion of light scattering/extinction data and for unraveling the effect of process variables (feedstock and oxidant flowrates) on the characteristics of carbon blacks. Wersborg et al. (29) modeled soot generation in premixed flames in terms of nucleation, condensation and coagulation treating the soot particles as spheres of uniform size. This model was in good agreement with their experimental data on soot from acetylene-oxygen flames when the coagulation rate was enhanced by a factor of 30. This enhancement was attributed to interparticle van der Waals forces.

Dobbins and Mulholland (39) modeled particle formation and coagulation in the free molecule regime in terms of the leading moments of the soot size distribution which was approximated by a lognormal function. They found that solutions to the soot population balance equation in which monodisperse or fixed width distributions were used, resulted in distorted particle concentrations along the flame axis. Megaridis and Dobbins (40) extended this model to bimodal distributions and accounted for surface growth reactions. Their results were in qualitative agreement with experimental studies of soot formation in premixed toluene/ethylene flames. Pratsinis (41) developed a similar model for particle nucleation, condensation (surface growth) and coagulation in the free molecule and continuum regimes approximating the aerosol size distribution by a unimodal lognormal function. He found that when surface growth and coagulation simultaneously take place the polydispersity of the product aerosol is lower than that obtained when only coagulation takes place.

Frenklach (42,43) also developed moment models for soot aerosol dynamics. He achieved closure of the moment equations by a series of approximations on the coagulation and condensation (surface growth) kernels without assigning a specific shape (self preserving or lognormal) to the soot size distribution. Frenklach et al. (44) developed detailed kinetic models predicting the inception rate of soot particles. Frenklach's (43) models were in good agreement with the model of Harris et al. (30). The latter model numerically solves the complete discrete population balance equation for 10,000 particle sizes. Although this is a brute force approach, its results are free of numerical artifacts arising from approximations of the shape of the soot size distribution. Of course, the major limitation of the model by Harris et al. (30) is its excessive computer time requirements: simulations of soot dynamics for 2 ms require 60 CPU minutes in a CRAY computer (43). As a result, discrete models are best suitable for the early stages of soot formation where there is substantial uncertainty on the distribution shape.

Kennedy (38) modeled free molecular soot dynamics in stagnation point diffusion flames utilizing a sectional representation of the aerosol size distribution. Using the fuel atom fraction, he calculated the flame temperature and species composition profiles without resorting to computation of the detailed flame kinetics. Kennedy (38) found that the calculated soot size distribution attained its self preserving form away from the flame front.

More recently the application of fractal theory is actively explored in modeling soot formation and growth. Samson et al. (45) found that an agglomeration model for fractal growth more closely reproduces the measured soot particle images than a diffusion limited aggregation model.

All models of soot dynamics employ the kinetic theory for coagulation and surface growth. There has been strong emphasis on the shape of soot size distribution from both measurement and modeling points of view. Moment models provide consistent results with the current understanding of soot generation at reasonable computation speed but at the expense of accuracy. The popularity of these models is justified since the approximations of the shape of the aerosol size distribution are insignificant compared to the ones associated with the estimation of soot inception rates, intermolecular forces, ionic effects and particle shape effects. Quantitative estimates for these phenomena are either obtained by comparison of model predictions with experimental data or simply neglected.

## CERAMIC POWDERS

Ceramic powders of submicron size are used to make pigments, catalytic substrates, silicone rubbers, semiconductors, superconductors and other advanced ceramics. These powders are made by wet chemistry, grinding and aerosol processes. Bowen (16) observed that production of ceramic powders by gas phase processes is most promising for large scale manufacture of advanced ceramics since it does not involve the tedious manufacturing processes of wet chemistry. Fine powders have been produced by aerosol processes in flames, shock tubes, laser beams, plasmas, furnaces, condensers-hydrolyzers and by expansion of supercritical solutions.

### Flame Reactors

In these systems particles are formed by chemical reactions from their precursor vapors. Usually small amounts of fuel (e.g. $CH_4$) are present in the feed to sustain the reaction. Production of rutile ($TiO_2$) by oxidation of $TiCl_4$ in flame reactors (the "chloride process") is one of the largest manufacturing aerosol processes. The commercial importance of the flame reactor has generated numerous patents on its design and operation (26) but very little is known about the fundamentals of rutile formation and growth. George et al. (46) investigated titania production in a premixed laminar flame reactor. They found that titania grew by coagulation to spherical particles having a self preserving size distribution.

Ulrich and his colleagues (47-49) extensively studied silica formation in laminar premixed and turbulent jet flame reactors. The silica was produced in aggregates of spherical primary particles. They suggested that the anomalous high viscosity of silica made fusion the controlling growth mechanism for primary particles and collision the controlling growth mechanism for silica aggregates. They employed sampling and microscopic analyses as well as in situ lazer characterization of the silica aggregates. They developed a sectional model for simulation of the silica aggregate aerosol dynamics in the free molecule and continuum regimes. Utilizing the initial surface area and temperature drop as adjustable parameters they obtained good agreement between their theory and data. They accounted for the irregular shape of silica aggregates by enhancing the collision rate in the free molecule regime by 20% (bulkiness factor, 1.2) based on the observations of Medalia and Heckman (37).

Nishida et al. (50,51) and Bolsaitis et al. (52) independently prepared ultrafine ($0.01 < d_p < 0.5 \mu m$) oxide powders by oxidation of metallic vapors. They used electric furnaces to generate high purity Mg and Zn vapors. Oxide powders were formed by injecting the metal vapors through a nozzle into the oxidation region of the furnace. Bolsaitis et al. (52) simulated ZnO powder production by the above process using the aerosol dynamics model of Gelbard and Seinfeld (53). They found, however, substantial disagreement between the computed ZnO particle size distributions and the measured ones by an electrical aerosol analyzer. This was attributed to the irregular (chainlike) shape of the ZnO powders.

Furnace Reactors

In these systems, particles are formed by chemical reactions of precursor gases in externally heated reaction vessels. Eversteijn (12) used these reactors to study the onset of silicon particle formation ("snow formation") during chemical vapor deposition of thin silicon films in epitaxial reactors.

Suyama and Kato (54,55) studied $TiO_2$ production by $TiCl_4$ oxidation in vertical and horizontal furnaces. They found by a series of microscopic and x-ray diffraction analyses that low concentrations of dopants (as low as 2% of inlet $TiCl_4$) have a pronounced effect on the crystallinity and size distribution of the product powder. This is an important result for industrial scale production of $TiO_2$ where dopants such as $AlCl_3$ and $SiCl_4$ are used for manufacture of rutile and anatase particles respectively (26).

Silicon nitride powders have been also made by gas phase reaction between $NH_3$ and $SiH_4$ in furnace reactors by Prochaska and Greskovich (56). They found that reactant stoichiometry and reactor temperature determined the color and crystallization temperature of the product powder by a series of microscopic and diffraction analyses.

Lay and Iya (57) developed a furnace reactor ('free space reactor') to make silicon particles by thermal decomposition of $SiH_4$ for low cost production of silicon for photovoltaic applications. Flagan and his coworkers (14,58,59) generated Si particles up to 10 $\mu m$ in size by thermally regulated $SiH_4$ decomposition and controlled Si nucleation and growth in two furnace aerosol reactors in series. They monitored product size distributions by conventional aerosol characterization instruments (optical particle counters, condensation nuclei counter and electrical aerosol analyzer) as well as microscopic (SEM and TEM) and x-ray diffraction analyses. By comparing reactor residence times to the characteristic collision times of seed Si particles with newly formed molecular Si clusters, Wu and Flagan (58) concluded that silicon growth took place by seed-cluster coagulation. The silicon particles have relatively narrow size distribution and Gregory et al. (60) made low density $Si_3N_4$ compacts by sintering and rapid nitridation of these powders. Hardness values of these compacts were comparable to commercial reaction-bounded silicon nitride composites.

Thermal decomposition of organometallic vapors in furnace reactors was investigated for production of ultrafine powders ($d_p < 0.01 \mu m$) in mixed (61) and premixed laminar (62,63) flows. Mazdiyasni et al. (61) made high purity, cubic $ZrO_2$ particles having a broad size distribution between 0.002 and 0.03 $\mu m$. These particles were highly sinterable and were used for deposition of thin zirconia films on hot substrates (refractory coatings). Okuyama et al. (63) investigated production of $Al_2O_3$, $TiO_2$ and $SiO_2$ particles by thermal decomposition of their organometallic (alkoxide) vapors in a furnace reactor. They measured the size, shape and composition characteristics of the product powders by real time aerosol instruments and microscopic/diffraction analyses at various reactor temperatures and inlet alkoxide vapor concentrations. Since in most experiments the decomposition reaction had not been completed in their furnace, the product particles were fine ($0.006 < d_p < 0.15$ μm) and had broad size distributions ($\sigma = 1.3 - 2$). They simulated their experiments using the computational scheme of Gelbard and Seinfeld (53) by assuming plug flow in the reactor and coagulation controlled particle growth. Fair agreement was obtained between theory and experiment when most of the inlet alkoxide vapor had been converted to oxide powder.

Laser Reactors

Nonoxide powders such as SiC, $Si_3N_4$ and Si have been made by heating precursor gases with a gas laser (64,65). This process achieves complete reactant mixing, reduces reactor wall deposits and results in high purity powders with fairly narrow size distribution since it confines the reaction to a small region where high temperatures and sort residence times prevail. Knudsen (66) prepared ultrafine ($0.01 < d_p < 0.1$ μm), equiaxed, loosely agglomerated $B_4C$ powders by $CO_2$ laser-driven pyrolysis of $BCl_3/H_2/CH_4$ mixtures. Flint et al. (65) determined particle size and concentration by real time, nonintrusive laser scattering/extinction measurements (35). They observed the particle concentration to decrease as a function of process residence time and attributed it to the change of the Si refractive index and to coagulation of newly formed Si or SiC particles. Knudsen (66) and Flint et al. (65) proposed nucleation and growth as the mechanisms for $B_4C$, Si and SiC particle growth. Rice (67) developed a 6-way cross laser reactor for manufacture of $SiC/Si_3N_4$ and $TiO_2$ powders from organometallic precursor vapors. Fairly polydisperse powders with significant neck formation were produced in his reactor. By controlling the reactant gas composition it was possible to control reaction temperature and powder stoichiometry and crystallinity.

Plasma Reactors

In these systems particles are formed by heating the reactant gases and promoting endothermic reactions by an electric plasma. Young and Pfender (68) reviewed generation of ceramic particles ($0.001 < d_p < 100$ μm) in a variety of plasma reactors during the last 40 years. Although temperature and residence time distributions are well understood, very little is known about particle formation and growth in plasmas. A variety of carbide, nitride and boride powders has been produced in RF plasmas (69). Precursor material can be fed in plasmas even in solid state. Recently, industry has been actively involved in developing plasmas for ceramic powder production and processing (8,70). Particle production in plasmas is carried out empirically since particle size, composition and crystallinity are usually controlled by trial and error procedures. Recently Gershick et al. (71) modeled particle production in RF plasmas by coagulation in one dimensional (plug) flow at constant cooling rate extending the computational scheme of Gelbard and Seinfeld (53). They simulated production of iron particles and found that the mean particle diameter was proportional to inlet iron vapor mole fraction and inversely proportional to plasma cooling rate. These computations were in fair agreement with the data of Yoshida and Akashi (72) and followed observations of other particle generation studies with RF plasmas (8).

Condenser-Hydrolyzers

Ceramic powders have been produced by low temperature aerosol processes such as formation of precursor organometallic droplets by condensation and subsequent in situ hydrolysis and conversion to hydroxide and oxide powders. Solid and porous titania and alumina powders have been produced by this process in laminar (73,74) and turbulent flows (24). These are fine ($0.1 < d_p < 1$ μm) amorphous powders with fairly narrow size distribution that can be transformed to crystalline by thermal treatment. Kodas et al. (24) showed that high volumes of alumina powders can be made by this process in turbulent flows at the expense of broadening of the particle size distribution. Pratsinis and his coworkers (21,75) developed numerical models simulating formation and growth of the organometallic aerosol. They investigated the effect of process variables on the average aerosol size, polydispersity and yield. Aerosol size and concentration characterizations have been carried out by optical particle counters (LAS-X, PMS) (24) or in situ light scattering instruments (76). This process is reasonably well understood since particle formation takes place by condensation onto seed nuclei or homogeneous nucleation while particle growth occurs by condensation.

## Expansion of Supercritical Fluids

Fabrication of ceramic powders and thin films can be accomplished by expansion of supercritical solutions at relatively low temperatures (less than 600°C) and high pressures (more than 100 atm). Solvents at supercritical conditions have the capacity to dissolve high amounts of solutes that are insoluble at normal conditions (77). Matson et al. (78) have shown that silica particles can be produced by expansion of supercritical aqueous - $SiO_2$ solutions. The product particles have narrow size distribution and range between 0.1 μm to 5 μm depending primarily on solute concentration and to lesser extent on pre-expansion temperature and pressure. X-ray diffraction analysis indicates that these particles are amorphous. Petersen et al. (79) postulated that particles produced by supercritical expansion of solutes are formed by nucleation with little influence by condensation and coagulation. This process has been also used to make $GeO_2$, $ZrO_2$ and organic particles and films.

## OPTICAL FIBERS

Aerosol processes are unique for production of materials with high purity. This is an attractive feature in lightguide fabrication technology. Optical fibers are made by a series of processes: fabrication of a preform glass rod by silica (and dopant) particle deposition, rod sintering, fiber drawing and coating (80). The key process, with respect to composition and purity of the product fiber, is the fabrication of the preform rod. The goal of this process is to make a preform rod having a prescribed radial distribution of refractive index at the maximum process yield.

Preforms are made by external and internal particle deposition processes. The outside vapor deposition (OVD) and vapor-phase axial deposition (VAD) processes are the external processes. They involve deposition of vapor and particles onto horizontal or vertical substrates from combustion of $SiCl_4$, $GeCl_4$ and other precursor gases. Glass particles are formed at the combustion zone of the gas burner and deposit as aggregates on the surface of the substrate (glass boule). Suda et al. (81) experimentally investigated the VAD process and found that the particle deposition depended largely on the Reynolds number of the flame stream. Similar results were obtained by Raychaudhuri and Biswas (82) in their VAD studies with a multiring burner. They emphasized, however, the importance of the gas velocity profile exiting the burner. Miller et al. (83) concluded that during OVD or VAD particle deposition at the vertical or horizontal substrate takes place by thermophoresis. The flame generated particles are too large for molecular diffusion and too small for impaction to be important. Rosner and Park (84) investigated the external processes for preform fabrication by analyzing transport of hot aerosols around cold wedges. They found that the particle deposition rate onto the wedge is higher than that predicted by classic thermophoretic calculations because high mass loadings of particles modify the gas stream velocity and temperature by an effect similar to 'massive suction' in single phase laminar boundary layer flows. Both Brownian and shear induced coagulation have a pronounced effect on the size distribution of the silica deposits (85).

Modified chemical vapor deposition (MCVD) and plasma chemical vapor deposition (PCVD) constitute the internal processes for preform manufacture. According to these processes, $O_2$, $SiCl_4$ and dopant vapors flow through a rotating quartz tube that is externally heated by a slowly, axially traversing, oxyhydrogen torch (or plasma). Inside the tube, the reactant gases are oxidized forming particles that either deposit to the tube walls or exit the tube with the process gases. Aside from oxidizing the reactants, the heat from the traversing torch also fuses the deposited particles forming a glassy layer in the interior of the substrate tube (13).

MacChesney et al. (86) invented the MCVD process for fabrication of lightguide preforms. Thermophoresis is the dominant mass transport mechanism as it has been proven experimentally (87) and theoretically (88,89). Rapid coagulation of the freshly formed oxide clusters in the torch area of the preform reduces the importance of molecular diffusion in favor of thermophoresis (90). Morse and Cipolla (91) found that the deposition efficiency (process yield) in MCVD can be substantially improved by increasing and maintaining the temperature gradient inside the preform tube by axial lazer heating of the process gases. Kim and Pratsinis (6,89) developed a detailed model of the MCVD process accounting for both gas phase kinetics and silica aerosol dynamics along the preform tube. They identified process conditions in which the MCVD deposition efficiency is limited by either mass transfer or chemical reaction.

A major challenge in multicomponent MCVD is to understand the relationship between glass particle size composition and process conditions. This is important because it has been observed that during sintering of the multicomponent particulate deposits small particles lose $GeO_2$ far more rapidly than large particles. The above relationship is also important for improvement of the current low process yields (~50%). As the fiber optics market becomes tighter this can be a critical issue from both economical and environmental viewpoints (especially for some of the toxic dopants).

## THIN FILMS FOR MICROELECTRONICS

Aerosol processes are also employed in fabrication of thin films for microelectronics (92). A common aerosol technology for thin film manufacture involves generation of an aerosol and its deposition onto a heated substrate. There, the droplets evaporate or react leaving the material that coats the substrate (93). This is the pyrosol process and its applications include manufacture of electrodes for solar cells, luminescent materials in CRT-screens and antireflective coatings (94). A major difficulty in this technology is the preparation of uniform droplets with diameter less than 1μm and the control of the physicochemical processes taking place near the deposition substrate.

Sputtering is another aerosol process for substrate coating. It involves material transport from a source to a substrate. The ejection of the source material is accomplished by bombardment of its surface with highly accelerated gas ions. Molecular clusters are ejected from the source and deposit by impaction onto the substrate thus forming metallic or ceramic films (95). Very little is known about the fundamentals of sputtering though various sputtering designs (diode, triode or magnetron) have been developed to control the deposition rate and reduce the charge buildup in the process equipment.

Etching patterns on thin films can be also accomplished by aerosol processes. Chen et al. (7) etched fine lines in silicon dioxide surfaces on silicon substrates by hydrofluoric acid aerosol jets. This aerosol process combines the high selectivity of wet etching with the fine line anisotropic etching of dry etching without the danger of radiation damage.

## CONCLUDING REMARKS

Aerosol processes are frequently utilized for manufacture of particulates with high purity, chemical homogeneity, small and uniform particle size and shape. Particulate commodities such as pigments and carbon blacks as well as materials for high technology applications such as optical fibers and advanced ceramics are currently made by aerosol processes on an industrial scale. Specific product requirements and engineering innovation has led to the development of a multitude of aerosol processes for material manufacture.

Historically, aerosol processes have phased out wet chemistry processes in industrial scale particulate operations. Typical examples are the dominance of the 'chloride' over the 'sulfate' process for manufacture of pigmentary titania and the success of dry scrubbing over the wet scrubbing processes for removal of pollutants from flue gases. The same trend is expected to develop for manufacture of ceramic powders that are currently made by sol-gel processes. Thus, there is strong interest for the development of efficient aerosol processes for manufacture of powders with controlled properties. These powders could be used for manufacture of highly tough ceramic parts, superconducting wires, packing material for chromatographic columns and inorganic membranes.

There is a wide range of progress stages for production of ceramic powders by aerosol processes. Production of titania and silica powders by flame reactors is a well established industrial aerosol operation. On the other hand, production of non-oxide powders (carbides, borides and nitrides) by aerosol processes is in a developmental stage. Various systems such as plasmas, lasers and expansion of supercritical solutions are currently being explored. Research in this area aims to invent new processes for production of powders with specific characteristics and to understand the fundamentals of existing aerosol processes for better monitoring and control of powder production. Invention of real time aerosol instruments and construction of comprehensive models for powder production constitute the key goals of this research.

Aerosol processes constitute one of the key operations in manufacture of optical waveguides. They determine the overall process yield and the distribution of refractive index across the fiber. There is good understanding of the physicochemical phenomena taking place during preform fabrication. Research in this area aims to develop quantitative models for optimal operation of existing industrial units and design of new highly efficient ones.

A paucity of general correlations between process variables and particulate characteristics, places serious limitations to the development, design, scale up and control of industrial aerosol operations. One of the encountered difficulties is the lack of standard instruments for real time particulate measurements. Conventional aerosol instruments such as electrical aerosol analyzers and optical particle counters find limited application for monitoring particulate production. These instruments require high dilution and transport of the sample away from its environment for measurement at room temperature.

Basic studies of soot formation in flames have led to the invention of nonintrusive optical techniques for real time characterization of highly concentrated aerosols. These techniques are already used in studies for production of ceramic powders and have high potential for monitoring industrial aerosol production. Research is needed to characterize generation of irregular particulates and to trace the early stages of particle formation by these instruments.

There are several models describing formation of particulates by moment, sectional and detailed solutions of the population balance equation. The model results serve two purposes. First, they relate process variables to particulate characteristics and second they are used to extract particulate size distributions from light scattering and extinction data. All models seem to perform reasonably well at least with respect to integral properties of the size distribution. Poor knowledge of specific physical phenomena (e.g. particle inception rates, Van der Waals forces etc.) does not allow, however, precise model predictions from first principles. Quantitative estimates for these phenomena are usually obtained by fitting model calculations through experimental data.

Although much remains to be learned about the fundamentals of aerosol formation and growth, enough is known to apply basic engineering principles to the design of industrial aerosol processes.

ACKNOWLEDGEMENT

The support from the National Science Foundation (grant CBT-8707144) and the DuPont De Nemours Co. (Dr. S.V.R. Mastrangelo) is gratefully acknowledged. This paper was presented at the 1988 AIChE Annual Meeting, Washington D.C., Nov 27-Dec 2.

LITERATURE CITED

1. Friedlander, S.K. Smoke, Dust and Haze, Wiley, New York, 1977.

2. Friedlander, S.K. Aerosol Sci. Technol. 1, 3 (1982).

3. Ulrich, G.D. Chem. Eng. News 62(32), 22 (1984).

4. Flagan, R.C., Seinfeld, J.H. Fundamentals of Air Pollution Engineering, Prentice Hall, Englewood Cliffs, New Jersey, 1988.

5. Walker, K.L., Geyling, F.T., Nagel, S.R. J. Amer. Ceram. Soc. 63, 552 (1980).

6. Kim, K.-S., Pratsinis, S.E. AIChE J. 34, 912 (1988).

7. Chen, Y.L., Brock, J.R., Trachtenberg, I. Appl. Phys. Lett. 51, 2203 (1987).

8. Sheppard, L.M. Adv. Mater. Proc. 4, 53 (1987).

9. Seinfeld, J.H. Atmospheric Chemistry and Physics of Air Pollution, Wiley, New York, 1986.

10. Rosner, D.E., Gunes, D., Nazih-Anous, N. Chem. Eng. Commun. 24, 275 (1983).

11. Nielsen, P.H. Villadsen, J. Chem. Eng. Sci. 41, 669 (1986).

12. Eversteijn, F.C., Philips Res. Repts 26, 134 (1971).

13. Nagel, S.R., MacChesney, J.B., Walker, K.L. IEEE J. Quant. Elec. QE-18, 459 (1982).

14. Alam, M.K., Flagan, R.C., Aerosol Sci. Technol. 5, 237 (1986).

15. Springer, G.S. Advances in Heat Transfer 14, 281 (1978).

16. Bowen, H.K. Mater. Sci. Eng. 44, 1 (1980).

17. Merrow, E.W. Chem. Eng. Prog. 81(5), 15 (1985).

18. Crump, J.C., Seinfeld, J.H. AIChE J. 26, 610 (1980).

19. Pratsinis, S.E., Friedlander, S.K., Pearlstein, A.J. AIChE J. 32, 177 (1986).

20. Pesthy, A.J., Flagan, R.C., Seinfeld, J.H. J. Colloid Interface Sci. 91, 525 (1983).

21. Pratsinis, S.E., Kodas, T.T., Sood, A. Ind. Eng. Chem. Res. 27, 105 (1988).

22. Pratsinis, S.E., Kodas, T.T., Dudukovic, M.P., Friedlander, S.K. Ind. Eng. Chem. Proc. Des. Dev. 25, 634 (1986).

23. Kodas, T.T., Friedlander, S.K., Pratsinis, S.E. Ind. Eng. Chem. Res. 26, 1999 (1987).

24. Kodas, T.T., Sood, A., Pratsinis, S.E. Powder Technol. 50, 47 (1987).

25. Medalia, A.I., Rivin, D., in Characterization of Powder Surfaces, (Parfitt, G.D., Sing, K.S.W) Academic, London, 1976.

26. Mezey, E.J., in Vapor Deposition (Powell, C.F., Oxley, J.H., Bocher, Jr. J.M. eds), Wiley New York, 1966.

27. Dannenberg, E.M. J. Inst. Rubber Ind. 5, 190 (1971).

28. Homann, K.H. 20th Symp. (Int.) on Combustion, 857 (1984).

29. Wersborg, B.L., Howard, J.B., Williams, G.C. 14th Symp. (Intl) on Combustion, 929 (1972).

30. Harris, S.J., Weiner, A.M., Ashcraft, C.C. Combust. Flame 64, 65 (1986).

31. Chung, S.-L., Katz, J.L. Combust. Flame 61, 271 (1985).

32. Santoro, R.J., Semerjian, H.G., Dobbins, R.A. Combust. Flame 51, 203 (1983)

33. Frenklach, M., Ramachandran, M.K., Matula, R.A. 20th Symp. (Int.) on Combustion, 871 (1984).

34. Santoro, R.J., Miller, J.H., Langmuir 3, 244 (1987).

35. D'Allesio, A., Di Lorenzo, A., Sarofim, A.F., Beretta, F., Masi, S., Venitozzi, C. 15th Symp. (Int.) on Combustion, 1427 (1974).

36. Dobbins, R.A., Megaridis, C.M. Langmuir 3, 254 (1987)

37. Medalia, A.I., Heckman, F.A. Carbon 7, 567 (1969).

38. Kennedy, I.M. 20th Symp. (Int.) on Combustion, 1095 (1984).

39. Dobbins, R.A., Mulholland, G.W. Combust. Sci. Technol. 40, 175 (1984).

40. Megaridis, C.M., Dobbins, R.A. Combust. Sci. Technol. in press (1988).

41. Pratsinis, S.E. J. Colloid Interface Sci. 124, 416 (1988).

42. Frenklach, M. J. Colloid Interface Sci. 108, 237 (1985).

43. Frenklach, M., Harris, S.J. J. Colloid Interface Sci. 118, 252 (1987).

44. Frenklach, M., Clary, D.W., Gardiner, Jr., W.C., Stein, S.E. 20th Symp. (Int.) in Combustion, 887 (1984b).

46. George, A.P., Murley, R.D., Place, E.R. Faraday Symposia 7, 63 (1973).

47. Ulrich, G.D., Combust. Sci. Technol. 4, 47 (1971).

48. Ulrich, G.D., Milnes, B.A., Subramanian, N.S. Combust. Sci. Technol. 14, 243 (1976).

49. Ulrich, G.D., Riehl, J.W. J. Colloid Interface Sci. 87, 257 (1982)

50. Nishida, A., Veki, A., Yoshida, K. Ceramic Powder Science (Advances in Ceramics) 21, 265 (1987).

51. Nishida, A., Yoshida, K., Igaroshi, H., Kobayashi, W. Ceramic Powder Science (Advances in Ceramics) 21, 271 (1987).

52. Bolsaitis, P.P., McCarthy, J.F., Mohiuddin, G., Elliot, J.F., Aerosol Sci. Technol. 6, 225 (1987).

53. Gelbard, F., Seinfeld, J.H. J. Colloid Interface Sci. 68, 363 (1979).

54. Suyama, Y., Kato, A. J. Amer. Ceram. Soc. 59, 146 (1976).

55. Suyama, Y., Kato, A. J. Amer. Ceram. Soc. 68, C-154 (1985).

56. Prochazka, S., Greskovich, C. Amer. Ceram. Soc. Bull. 57, 579 (1978).

57. Lay, J.R., Iya, S.K. 15th IEEE Photovoltaic Specialists Conference, 565 (1981).

58. Wu, J.J., Flagan, R.C. J. Appl. Phys. 61, 1365 (1987).

59. Wu, J.J., Nguyen, H.V., Flagan, R.C. Langmuir 3, 266 (1987).

60. Gregory, O.J., Lee, S.-B., Flagan, R.C. J. Amer. Ceram. Soc. 70, C-52 (1987)

61. Mazdiyasni, K.S., Lynch, C.T., Smith, J.S. J. Amer. Ceram. Soc. 48, 372 (1965).

62. Komiyama, H., Kanai, T., Inoue, H. Chem. Lett. 1283 (1984).

63. Okuyama, K., Kousaka Y., Tohge N., Yamamoto, S., Wu, J.J., Flagan, R.C., Seinfeld, J.H. AIChE J. 32, 2010 (1986)

64. Cannon, W.R., Danforth, S.C., Flint, J.H. Haggerty, J.S., Marra, R.A., J. Amer. ceram. Soc. 65, 324 (1982).

65. Flint, J.H., Marra, R.A., Haggerty, J.S. Aerosol Sci. Technol 5, 249 (1986).

66. Knudsen, A.K. in Ceramic Powder Science (Adv. in Ceramics) 21, 237 (1987).

67. Rice, G.W. Ceramic Powder Science (Advances in Ceramics) 21, 229 (1987).

68. Young, R.M., Pfender, E., Plasma Chem. Plasma Process. 5, 1 (1985).

69. Vogt, G.J., Phillips, D.S., Taylor, T.N. Ceramic Powder Science (Adv. in Ceramics) 21, 203 (1987).

70. Smith, F.N., Liu, J., Becker, A. Meyer, T.N. Aerosols '87 Pilat, M.J., Davis, E.J. eds., American Association for Aerosol Research, 143 (1987).

71. Girshick, S.L., Chiu, C.-P., McMurry, P.H. Plasma Chem. Plasma Process. 8 145 (1988).

72. Yoshida, T., Akashi, K. Trans. Japan Instit. Metals 22, 371 (1981)

73. Visca, M., Matijevic, E. J. Colloid Interface Sci. 68, 308 (1979).

74. Ingebrethsen, B.J., Matijevic, E., Partch, R.E. J. Colloid Interface Sci. 95, 228 (1983).

75. Phanse, G.M., Pratsinis, S.E. Aerosol Sci. Technol. in press, 1989.

76. Kerker, M. The Scattering of Light, Academic, New York, 1969.

77. Paulaitis, M.E., Krukonis, V.J., Kurknik, R.T., Reid, R.C. Rev. Chem. Eng. 1, 179 (1983).

78. Matson, D.W., Peterson, R.C., Smith, R.D. Ad. Ceram. Mater. 1, 242 (1986).

79. Petersen, R.C., Matson, D.W., Smith, R.D. J. Amer. Chem. Soc. 108, 2100 (1986).

80. Rowell, J.M. Scientific American, October, 147 (1986).

81. Suda, H., Sudo, S., Nakahara, M. Fiber and Integrated Optics 4, 427 (1983).

82. Raychaudhuri, S., Biswas, D.R. J. Amer. Ceram. Soc. 67, C-57 (1984).

83. Miller, T.J., Potkay, E., Yuen, M.J. AIChE Symp. Series 83(258), 1 (1987).

84. Rosner, D.E., Park, H.M. Chem. Eng. Sci., 43, 2689 (1988).

85. Park, H.M., Rosner, D.E. Chem. Eng. Sci. 44, 603 (1989).

86. MacChesney, J.B., O'Connor, P.B. and Presby, H.M. Proc. IEEE 62, 1278 (1974).

87. Simpkins, P.G., Greenberg-Kosinski, S., MacChesney, J.B. J. Appl. Phys. 50, 5676 (1979).

88. Walker, K.L., Homsy, G.M., Geyling, F.T. J. Colloid Interface Sci. 69, 138 (1979).

89. Kim, K.-S., Pratsinis, S.E. Chem. Eng. Sci. accepted for publication, 1989.

90. Pratsinis, S.E., Kim, K.-S. J. Aerosol Sci. 20, 101 (1989).

91. Morse, T.F. and Cipolla, J.W. J. Colloid Interface Sci. 97, 137 (1984).

92. Kodas, T.T., Baum, T.H., Comita, P.B. J. Appl. Phys. 62, 281 (1987).

93. Blandenet, G., Court, M., Lagarde, Y. Thin Solid Films 77, 81 (1981).

94. Lambeck, P.V., Hildernik, L., Popma, T.J.A. Aerosols: Formation and Reactivity, 2nd Int. Aerosol Conf. Berlin, Pergamon, 972 (1986).

95. Wachtman, J.B., Haber, R.A. Chem. Eng. Prog. 82(1), 39 (1986).

# SELECTED APPLICATIONS OF ULTRA-RAPID FLUIDIZED (URF) REACTORS: ULTRAPYROLYSIS OF HEAVY OILS AND ULTRA-RAPID CATALYTIC CRACKING

A.L. Vogiatzis, C.L. Briens and M.A. Bergougnou ∎ The University of Western Ontario, Department of Chemical and Biochemical Engineering, Faculty of Engineering Science, London, Ontario, Canada, N6A 5B9

An experimental study was conducted at high reaction temperatures (500–900 °C) and short reaction residence times (70–500 ms) using a bitumen feedstock from Cold Lake (Alberta, Canada). The experiments were conducted on the ultrapyrolysis mini-pilot plant facility at the University of Western Ontario. The results indicated that the ultrapyrolysis process was effective in converting the bitumen feed into a favourable product distribution which included valuable chemical intermediates. The total gas product formed was selectively olefinic; more than 18 wt% of the bitumen feedstock was converted to ethylene at 900 °C and 331 ms, when the conversion of bitumen to total gas was 51.23 wt%. Preliminary analysis of the liquid products indicated reductions in viscosity, sulphur and nitrogen contents, and metals content. Solid residue formation constituted less than 14 wt% of the feed in all experiments conducted. Preliminary catalytic experiments were also conducted on a vacuum gas oil feedstock, and the results showed the potential of operating at the conditions of the ultrapyrolysis process (high temperatures, short residence times) for increased conversions and improved selectivities.

It is of the greatest urgency now, to develop processes to convert abundant and cheap carbonaceous fuels such as biomass, high volatile coals, petroleum residuals, and tar sand bitumen into <u>high-value added feedstocks</u> which can be accepted by conventional refineries to produce the transportation fuels and petrochemicals to which we are accustomed. What is, indeed, sorely needed is a greater feedstock versatility than the industry now has and resulting increased resilience for the economy. For instance, even right now, we need new processes to deal with the bottom of the barrel in refineries, which cannot be simply burnt because of $SO_x$ and metals pollution problems.

Processes for the hydroliquefaction of heavy carbonaceous fuels will be, for a long time, too expensive, because of the high cost of hydrogen and of high-pressure equipment. It is, thus, better to reject carbon rather than to add hydrogen to heavy carbonaceous feedstocks. Ultra-rapid pyrolysis (ULTRAPYROLYSIS) allows the recovery of valuable volatiles from the latter. It requires very efficient mixing, at <u>high temperatures</u>, of the cold feedstock with the reactor contents and very fast separation of the vapour products from the remaining solids, the products being subsequently sent to quench to prevent degradation of the chemical intermediates obtained.

Conventional <u>slow</u> fluidized beds are not easily adaptable to high-temperature fast reactions at a commercial scale. Relaxation times for mixing of the particulate feedstock with bed solids are of the order of 2-40 seconds; these times are too long for reactions taking place on a millisecond scale. Vapour product residence times are also of the order of seconds, which is also far too long to prevent thermal degradation of the products. Even the commercial upflow fast fluidized bed has solid mixing times of the order of one second or more (in addition, a lot of solids and entrained gas reflux downwards along the walls). A new type of reactor is thus needed for fast, high-temperature, pyrolytic reactions, which require, for optimal results, short contact times (in the hundreds of milliseconds), plug flow and fast separation.

<u>Jet Impact Reactors</u>.

All reactors involve the jetting of the feedstock into the reactor contents. How this jetting is done is not important for slow reactions. On the contrary, it is of the greatest importance for fast, complex, highly branched reactions.

The large particle hold-ups of slow and fast fluidized beds are an impediment to feedstock jet dissipation. Hence, a more viable approach should be taken, which includes impacting all jets one onto another, using an all-jet reactor. Instead of the energy of mixing being distributed across the bed, it is concentrated at the point where the jets impact one another. The concept is analogous to the transistor in electronics. In old-type vacuum tubes, the space charge of the electron holdup between the anode and the cathode created problems. The transistor dispensed with the large volumes of vacuum tubes and kept only the junctions. In a similar fashion, the jet impact reactor keeps only the jets.

JET IMPACT reactors are, thus, major new high-performance reactors which have many potential applications in a great variety of processing fields. They essentially impact jets of a hot (around 800-1000°C) heat carrier or thermofor (a gas; or an atomized liquid; or fine solid particulates; catalytic or not) onto a jet of relatively cold feedstock (a gas; or atomized liquid; or fine solid particulates) to effect: (i) rapid mixing of the hot thermofor with the relatively cold feedstock (typically in less than 30 milliseconds); (ii) thermal shock of the feedstock. Multiphase shock waves can be induced in the various jets for more intense mixing at the point of jet impact. Under sonic and thermal shocks, solid particulate feedstocks "ablate". The vapours thus created break down into free radicals which, then, recombine into useful chemical intermediates. The mixed phases continue to react, as needed, in a down-flow transported reactor, in a plug-flow fashion, for typically a few hundred milliseconds. The vapour product is then separated quickly from the solid particulates, if such have been used, in a fast uniflow cyclone (1) before being sent to quench, which keeps chemical intermediates from decomposing. The particulate thermofor is sent to a reheat vessel (where coke is burnt off) before being recycled to the jet impact reactor.

## Mini-Pilot Plant

A mini-pilot plant has been built on the above concept (0.5 kg/hr. feedstock) and is now pyrolyzing a variety of heavy carbonaceous feedstocks (cellulose, biomass, heavy oils, asphalts, etc.). The rationale for ultra-rapid pyrolysis (Ultrapyrolysis) lies in the fact that the molecular structure of natural, heavy, carbonaceous fuels is not strongly cross-linked. If appropriately hit by a high-temperature thermal shock (fast pyrolysis), these feedstocks will shatter and disintegrate into smaller reactive molecular fragments before the structure has time to become cross-linked and refractory. On the contrary, slow pyrolysis would leave enough time for the structure to become cross-linked and refractory thus leading to a lot of worthless tar and coke. Another advantage of ultrapyrolysis is the compactness of the equipment due to the fast kinetics at high temperatures. Because of feedstock versatility (any carbonaceous fuel or residual feedstock), ultrapyrolysis has great promise to be a cost effective process in the near future.

The mini-pilot plant can also operate in a catalytic mode, the catalyst being the thermofor. Work is being planned on advanced catalytic cracking schemes, on direct oxidation of methane to methanol and on coupled oxidative dehydrogenation of methane to ethylene.

In the present mini-pilot plant (Figure 1), the thermofor enters the preheater coil and is heated to a sufficient temperature above the reaction temperature, as dictated by the heat balance. Reactor temperatures can be set in the range of 500 to 950°C. The fast initial temperature rise time is achieved through direct contact of the atomized liquid feedstock (preheated to approximately 200°C to reduce its viscosity) with a heat transfer medium (nitrogen gas to which 100 $\mu$m sand particles may be added) in the thermovortactor. The thermovortactor has two opposing tangential inlets. One tangential stream effectively destroys the momentum of the other causing severe turbulence. The liquid feedstocks are then injected from the top of the thermovortactor through an air cooled tube into the turbulent region where mixing occurs within 30 milliseconds. Pyrolysis continues in the reaction section, which is simply a tube kept under isothermal conditions. The reaction continues for a predetermined residence time and its products are then quenched by a stream of cold nitrogen as they enter the cryovortactor. At this point the mixture is cooled to less than 300°C. The product stream proceeds to a water-cooled cyclonic condenser, a cold trap, an electrostatic precipitator and a porous metal filter cartridge, so that only the non-condensible products reach the gas collection bags.

Experimental results have been obtained, using the approach described above, on the ultrapyrolysis mini-pilot plant facility at the University of Western Ontario. The experimental study was initiated using Cold Lake bitumen (10.2 API gravity; MCR 13.1 wt%; 4.4 wt% sulphur; 0.45 wt% nitrogen; 190 wppm vanadium; 77 wppm nickel; and 6390 cSt/40°C viscosity) to determine the ability for upgrading. The initial results at selected temperatures and residence times were encouraging (2). The results presented in this paper complete the study of the Cold Lake bitumen over wider ranges of temperature and residence time.

## RESULTS AND DISCUSSION

A study was conducted on the pyrolysis of a Cold Lake bitumen feedstock, over a range of temperatures (500-900°C) and residence times (70-500 ms); the effect of these two experimental parameters on the overall yields of the three product fractions (gas, liquid, solid residue) was determined. In addition, the individual gas component yields were obtained from gas chromatography analyses (Carle GC, model 111-H, 197-A). Mass balances were obtained from the three product fractions collected, and typically over 92 wt% of the bitumen fed was accounted for in all cases. All yields in this paper are reported in wt%, and were computed on the basis of the mass of bitumen feedstock pyrolyzed in a given experiment.

### Effect of Residence Time

Table 1 summarizes the results obtained at various residence times for two extreme temperatures in this study, 605°C and 900°C.

At 605°C, the total gas yield increases linearly with increasing residence time, and a slope of 0.81 wt%/100 ms is obtained. This gives a 3.3 wt% increase in the total gas yield over the residence time range studied, from 103 to 500 ms. The gas component yields at 605°C are shown in Table 1, and plotted in Figure 2 as a function of residence time. The yields of ethylene, propylene, methane, ethane, propane, and hydrogen sulfide increase linearly with residence time, which indicates that there is no secondary cracking of these components. The gas component with the highest yield at 605°C is the $C_5$-$C_6$ fraction, reaching yields of up to 2.8 wt% at 500 ms. As shown in Figure 2, however, the increase in $C_5$-$C_6$ yield with residence time is not linear. This may be due to some secondary cracking of the $C_5$-$C_6$ at longer residence times, which would yield the lighter hydrocarbon products, and $C_4$-olefins after about 300 ms.

The effect of residence time on the total gas yield, as well as on the gas product distribution is largely affected by temperature. At lower temperatures (500-650°C), the total gas yield increases linearly with residence time (for residence times between 100-500 ms). Within the 500-650°C temperature range, only the slope of the line varies, increasing at higher temperatures. For example, the slope for the total gas yield as a function of residence time line is 0.81 wt%/100 ms at 605°C, and 4.7 wt%/100 ms at 650°C.

At higher temperatures (850, 900°C), the total gas yield does not increase linearly over the entire range of residence times (70-400 ms). At the longer residence times (typically 250-500 ms), the slope for the total gas yield vs. residence time curve begins to decrease, as secondary cracking reactions become significant. Table 1 and Figure 3 show experimental results at 900°C. At residence times above 124 ms, the yields of $C_4$-olefins, propylene, and ethane decrease, and the yields of hydrogen and acetylene increase. In addition, the slopes of the ethylene and $C_5$-$C_6$ yield curves decrease as a function of residence time. Gas components which are less refractory, such as ethane, $C_4$-olefins, and $C_5$-$C_6$, undergo secondary cracking more readily than ethylene, which is more refractory. Methane is the most refractory component, and requires higher temperatures (> 1000°C) before any significant cracking is observed at these residence times (<500 ms). This explains why the methane yield continues to increase.

One advantage for operating at long residence times and high temperatures, is that more sulfur is removed from the liquids as hydrogen sulfide gas. At 900°C and 331 ms, nearly 2.6 wt% hydrogen sulfide gas forms. This represents more than 50% removal of the sulfur contained in the Cold Lake bitumen feed.

At higher reaction temperatures, approaching 900°C, optimal yields and product distributions are obtained at shorter reaction residence times (<150 ms). For example, to maximize the ethylene concentration in the product gas, and to produce a gas richer in olefins, very short residence times (about 70 ms) are required at the high reaction temperature of 900°C. This also minimizes, methane, hydrogen and acetylene yields.

The effect of residence time on the gas product distribution, thus, depends on the temperature. At higher temperatures, the thermal cracking kinetics are faster, and secondary cracking reactions to low value products (hydrogen, acetylene, methane) are initiated sooner.

### Effect of Temperature

Table 2 summarizes the results obtained at various temperatures (600-900°C) for short (120 ± 17 ms) and for long (390 ± 17 ms) residence times.

At short residence times, the total gas yield increases linearly with temperature. Figure 4 shows the individual gas component yields as a function of temperature, for short residence times. The ethylene yield increases with temperature, and represents the highest olefinic gas component yield, reaching about 18 wt% at 900°C. For short residence times, the yield of 1,3-butadiene increases continuously with temperature, reaching about 3.0 wt% at 900°C. Evidence of secondary cracking reactions is given by the decrease in $C_5$-$C_6$ and $C_4$-olefin yields, the increase in the methane yield and the formation of hydrogen and acetylene at the higher temperatures (>780°C), as shown in Table 2.

At long residence times, the total gas yield increases linearly with temperature, until about 800°C. Thereafter, secondary cracking reactions become significant, and the slope of the line begins to fall off. Figure 5 shows the individual gas component yields as a function of temperature, for long residence times. The ethylene and methane yields continue to increase at high temperatures. At 900°C and long residence time, the ethylene yield is about 19 wt%, which compares well with the 18 wt% ethylene obtained at 900°C and short residence time. The methane yield at 900°C increases from 7.7 wt% at the short residence time (Figure 4), to 10.8 wt% at the long residence time (Figure 5). Above 800°C, at long residence times, secondary cracking of the $C_3$-$C_6$ gas components becomes significant since the yields of $C_5$-$C_6$, propylene, 1,3-butadiene and $C_4$-olefins decrease. In addition, as shown in Table 2, hydrogen and acetylene yields increase at long residence times and high temperatures. Between 800 and 900°C, the slope of the total gas yield vs temperature curve quickly falls off. At these high temperatures, secondary cracking reactions lead to more solid residue formation and higher yields of hydrogen. This suggests that carbon is being rejected as a result of the severe reaction conditions, and a lower total gas yield is obtained.

A more favourable product distribution, giving lower solid residue yields and higher concentration of olefins in the total gas yield, is obtained at <u>short</u> residence times (< 150 ms) at high temperatures (> 800°C). Figure 6 shows the yield of olefins as a function of reaction temperature, at short and long residence times. The yield of olefins decreases at long residence times when the temperature is above 800°C, which reflects the secondary cracking of the $C_3$-$C_4$ olefins; the ethylene yield, however, continues to increase (see also Table 2), and ethylene is the major gas component. For commercial application, the liquids may be recycled, thus increasing the yields of olefinic compounds. This increase would be more pronounced at shorter residence times. Figure 7 shows the variation in the $C_5$-$C_6$ yield and hydrogen sulfide yield as a function of temperature. The $C_5$-$C_6$ fraction increases until about 750°C, for both short and long residence times, after which it undergoes secondary cracking; this is due to the faster reaction kinetics at the higher temperatures. The hydrogen sulfide yield, however, continues to increase even at the most extreme temperatures, and the maximum yield (about 2.6 wt%) is obtained at long residence times and high temperatures. To produce a more olefinic product gas distribution, and to achieve a moderate removal of sulfur as hydrogen sulfide gas, reaction should be carried out at short residence time (50-80 ms) and high temperature (about 900°C), with the liquid products being recycled.

Tables 1 and 2 summarize a representative portion of the experimental data obtained for Cold Lake bitumen pyrolysis for the temperature ranging from 500-900°C and residence times ranging from 70-500 ms. Figure 8 was developed from all of the experimental data obtained at these conditions. This plot shows the three product fractions collected from each experiment for mass balance considerations: i.e. total gas yield, total liquid yield and solid residue (in all of the experiments from which this plot was developed, the mass of collected products was more than 92 wt% of the feed). This plot, thus, summarizes the expected yields of the three product fractions over the range of temperatures and residence times studied. The solid curves for each product fraction indicate an intermediate residence time which decreases with temperature; that is, as the temperature in the reactor increases from 500 to 900°C, this intermediate residence time decreases from 250 to 200 ms. The dotted lines around the total gas and total liquid curves indicate the variation of each fraction with residence time, for a given temperature. The effect of residence time is more pronounced at the higher temperatures. The solid residue does not vary to any appreciable extent for a given temperature; when the total gas yield increases at longer residence times (for a given temperature), the total liquid produced decreases by a corresponding amount, and the solid residue yield is not greatly changed. The total liquid yield decreases at higher temperatures. From Figure 8, then, the overall composition of the product fractions can be estimated over the entire temperature range from 500-900°C and residence time range from 70-500 ms, for Cold Lake bitumen pyrolysis.

Figure 9 summarizes the overall composition of the total gas as a function of temperature. This plot can be applied at residence times up to 500 ms. The yield of any gas component, as a mass fraction of the total gas, does not vary appreciably with residence time for any given temperature in this study. Hence, Figure 9 is applicable at all residence times from 70-500 ms. Combined with Figure 8, the total gas yield and its composition can be estimated at any given temperature and residence time within the scope of this study.

### Catalytic Experiments

Recently, preliminary experiments have been conducted on a vacuum gas oil feedstock. Experiments were conducted at 550°C and about 550 ms. Super Nova-D catalyst was used with a catalyst to oil ratio of 6. Further experiments are currently underway at higher temperatures and short residence times. The fast uniform mixing of the oil feedstock with the hot catalyst (achieved in the mixing zone of the ultrapyrolysis reactor by the impinging jets), as well as the near plug flow behaviour of the gases and solids in the reactor should result in increased conversions and improved selectivities.

### CONCLUSIONS

The experimental work completed on the ultra-rapid pyrolysis of Cold Lake bitumen indicates the potential of this process for upgrading heavy oil feedstocks. The total gas produced is selectively olefinic at high temperatures, with more than 18 wt% of the bitumen being converted to ethylene at the most extreme experimental conditions of this study. Analyses of the liquid products are also expected to show a favourable degree of upgrading; preliminary analyses indicate that the liquid is of much lower viscosity than the feed, and has a lower sulphur, nitrogen, and metals content. The results are encouraging with respect to process economics and selectivity, as well as flexibility. The preliminary catalytic work on the vacuum gas oil indicates the potential for development of the ultra-rapid pyrolysis (ultrapyrolysis) process in the catalytic cracking mode. Furthermore, catalytic studies will provide important information on the advantages (increased conversion, improved selectivities, better process economics) of the ultrapyrolysis process over conventional FCC processes.

### ACKNOWLEDGEMENTS

This work was funded by grants from Imperial Oil Limited and the Department of Energy, Mines and Resources (EMR). Their help made this work possible and is gratefully acknowledged here.

### REFERENCES

1. Sumner, R.J., C.L. Briens, M.A. Bergougnou, Can. J. Chem. Eng., 65, 470-475 (1987).

2. A.L. Vogiatzis, S. Afara, C.L. Briens and M.A. Bergougnou, "Ultrapyrolysis of Cold Lake Bitumen", in the Proceedings of the Second International Conference on Circulating Fluidized Beds, Compiegne, France, March 14-18, Permagon Press, pp.483-490 (1988).

Table 1. Variation of the total gas yield and gas component yields with RESIDENCE TIME, at 605 and 900 °C; for Cold Lake bitumen feedstock.

| REACTION RESIDENCE TIME (ms) | REACTION TEMPERATURE (°C) | MASS BALANCE (wt%) | PRODUCT YIELDS (wt% of feed) | | | GAS COMPONENT YIELDS (wt% of feed) | | | | | | | | | | | |
|---|---|---|---|---|---|---|---|---|---|---|---|---|---|---|---|---|---|
| | | | | | | | | TOTAL OLEFINS | | | | | TOTAL PARAFFINS | | | | |
| | | | Gas | Liquid | Solid* | $H_2$ | $C_2H_2$ | $C_2H_4$ | $C_3H_6$ | $C_3H_4$ | 1,3-Butadiene | $C_4^=$ | $CH_4$ | $C_2H_6$ | $C_3H_8$ | $C_5$-$C_6$ | $H_2S$ |
| 103 | 605 | 94.7 | 3.05 | 85.4 | 6.2 | - | - | 0.24 | 0.24 | - | - | - | 0.4 | 0.25 | 0.13 | 1.5 | 0.29 |
| 160 | 605 | 96.2 | 3.76 | 80.7 | 11.7 | - | - | 0.45 | 0.45 | - | - | - | 0.45 | 0.31 | 0.15 | 1.6 | 0.35 |
| 316 | 605 | 98.0 | 5.41 | 82.9 | 9.7 | - | - | 0.60 | 0.57 | - | - | 0.13 | 0.52 | 0.36 | 0.16 | 2.7 | 0.37 |
| 411 | 605 | 98.2 | 5.69 | 86.6 | 5.9 | - | - | 0.58 | 0.57 | - | - | 0.27 | 0.62 | 0.4 | 0.18 | 2.6 | 0.47 |
| 500 | 605 | 99.3 | 6.32 | 86.9 | 6.1 | - | - | 0.72 | 0.71 | - | - | 0.38 | 0.67 | 0.38 | 0.2 | 2.8 | 0.46 |
| 73 | 900 | 92.6 | 36.04 | 46.4 | 10.2 | 0.25 | 0.95 | 16.4 | 3.0 | 0.37 | 2.0 | - | 6.0 | 1.4 | - | 4.96 | 0.71 |
| 124 | 900 | 92.4 | 48.52 | 39.3 | 4.6 | 0.55 | 1.14 | 17.89 | 4.3 | 0.52 | 3.0 | 1.22 | 7.7 | 1.1 | - | 9.7 | 1.40 |
| 331 | 900 | 100.0 | 51.23 | 34.8 | 14.0 | 1.04 | 1.22 | 18.65 | 2.96 | 0.28 | 2.0 | 0.56 | 10.8 | 0.7 | 1.05 | 9.4 | 2.58 |

* Solid residue collected in mass balance procedure.

Table 2. Variation of the total gas yield and gas component yields with TEMPERATURE, at short (120 ± 17 ms) and long (391 ± 60 ms) residence times; for Cold Lake bitumen feedstock.

| REACTION TEMPERATURE (°C) | REACTION RESIDENCE TIME (ms) | MASS BALANCE (wt%) | PRODUCT YIELDS (wt% of feed) | | | GAS COMPONENT YIELDS (wt% of feed) | | | | | | | | | | | |
|---|---|---|---|---|---|---|---|---|---|---|---|---|---|---|---|---|---|
| | | | | | | | | TOTAL OLEFINS | | | | | TOTAL PARAFFINS | | | | |
| | | | Gas | Liquid | Solid* | $H_2$ | $C_2H_2$ | $C_2H_4$ | $C_3H_6$ | $C_3H_4$ | 1,3-Butadiene | $C_4^=$ | $CH_4$ | $C_2H_6$ | $C_3H_8$ | $C_4^-$ | $C_5$-$C_6$ | $H_2S$ |
| 605 | 103 | 94.7 | 3.05 | 85.4 | 6.2 | - | - | 0.24 | 0.24 | - | - | - | 0.40 | 0.25 | 0.13 | - | 1.5 | 0.29 |
| 640 | 137 | 98.8 | 8.86 | 79.8 | 10.1 | - | - | 0.84 | 0.75 | - | - | - | 0.72 | 0.41 | 0.20 | - | 5.5 | 0.44 |
| 710 | 137 | - | 26.20 | - | - | - | - | 4.12 | 3.10 | - | 0.87 | 1.68 | 1.90 | 0.80 | 0.28 | - | 12.7 | 0.75 |
| 850 | 131 | - | 36.90 | - | - | 0.34 | 0.29 | 14.40 | 3.60 | - | 1.98 | 1.44 | 6.80 | 1.60 | - | - | 5.7 | 0.75 |
| 900 | 124 | 92.4 | 48.52 | 39.3 | 4.6 | 0.55 | 1.14 | 17.89 | 4.3 | 0.52 | 3.0 | 1.22 | 7.70 | 1.10 | - | - | 9.7 | 1.40 |
| 600 | 439 | 94.9 | 6.30 | 78.6 | 10.0 | - | - | 0.67 | 0.69 | - | - | 0.27 | 0.72 | 0.47 | 0.20 | 0.12 | 2.6 | 0.5 |
| 662 | 428 | 97.4 | 21.85 | 62.3 | 13.2 | - | - | 3.55 | 2.75 | - | 0.73 | 1.52 | 1.60 | 0.67 | 0.17 | - | 10.28 | 0.58 |
| 700 | 451 | 98.7 | 29.26 | 60.3 | 9.1 | 0.09 | - | 4.94 | 4.3 | - | 1.28 | 2.57 | 3.00 | 1.54 | 0.37 | 0.13 | 9.92 | 1.12 |
| 750 | 422 | 99.6 | 42.93 | 47.6 | 9.1 | 0.25 | - | 8.73 | 6.7 | - | 2.54 | 3.56 | 5.30 | 2.10 | 0.44 | 0.11 | 11.8 | 1.4 |
| 800 | 373 | - | 51.33 | - | - | 0.49 | 0.25 | 14.00 | 7.4 | - | 4.25 | 1.94 | 6.80 | 1.30 | 0.50 | 0.30 | 12.4 | 1.7 |
| 900 | 331 | 100.0 | 51.23 | 34.4 | 14.4 | 1.04 | 1.22 | 18.65 | 2.96 | 0.28 | 2.00 | 0.56 | 10.80 | 0.70 | 1.05 | - | 9.4 | 2.58 |

* Solid residue collected in mass balance procedure.

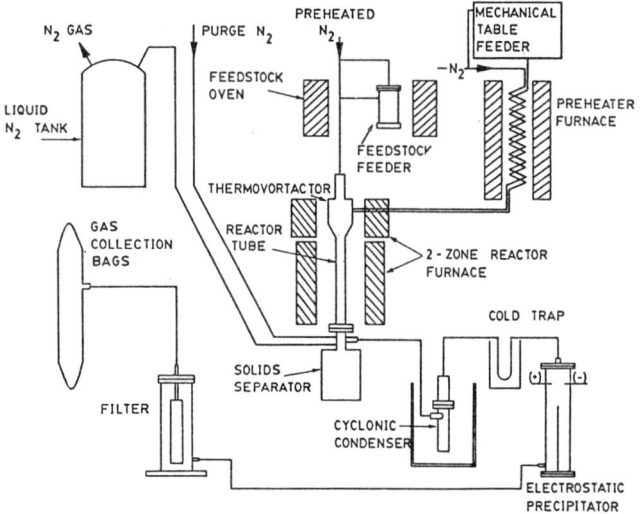

Figure 1. Ultrapyrolysis flow scheme.

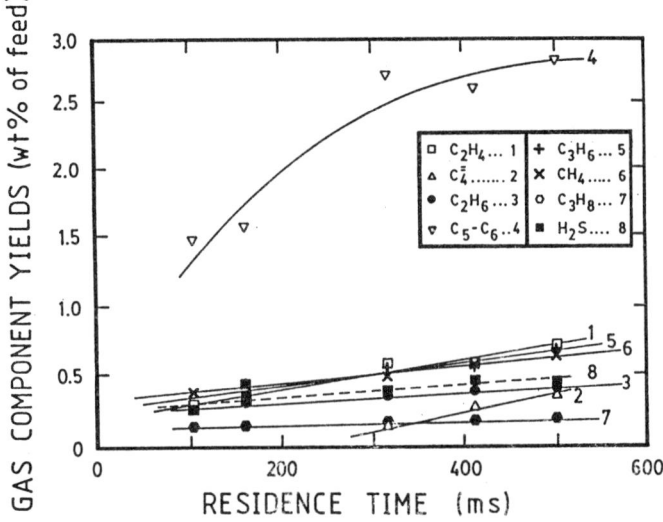

Figure 2. Plot of gas component yields as a function of reaction residence time, at 605 °C.

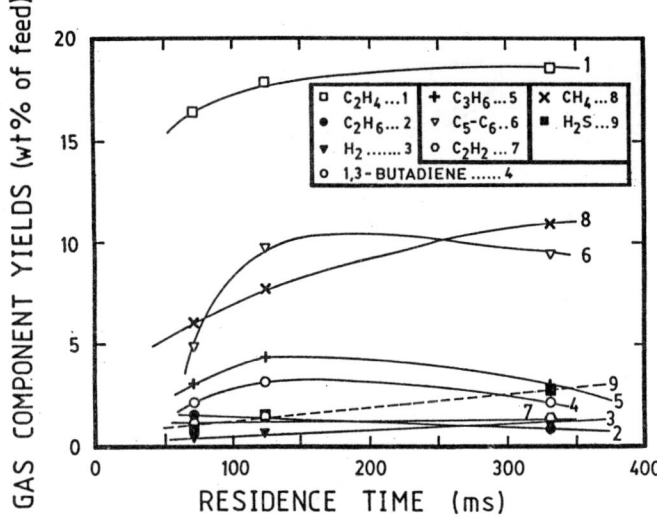

Figure 3. Plot of gas component yields as a function of reaction residence time, at 900 °C.

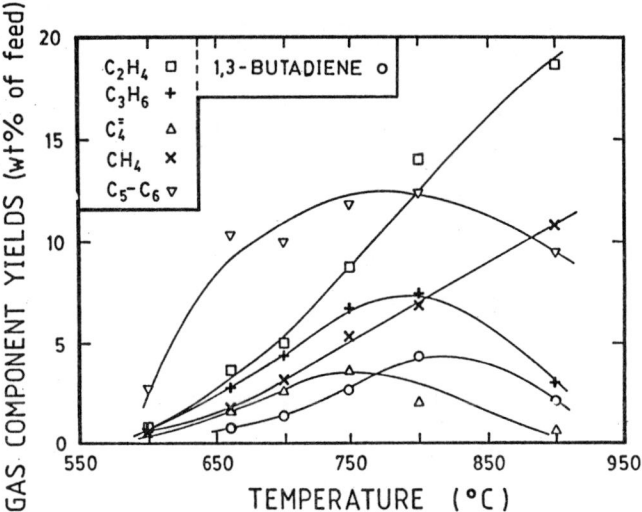

Figure 5. Plot of the major gas component yields as a function of temperature, at 391 ± 60 ms.

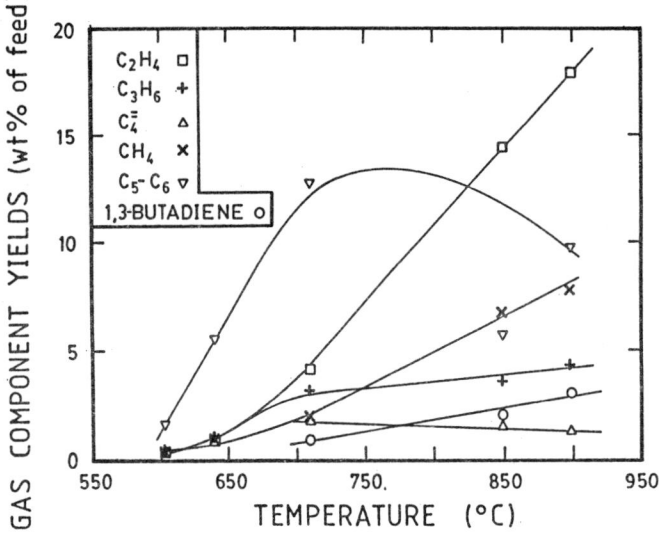

Figure 4. Plot of the major gas component yields as a function of temperature, at 120 ± 17 ms.

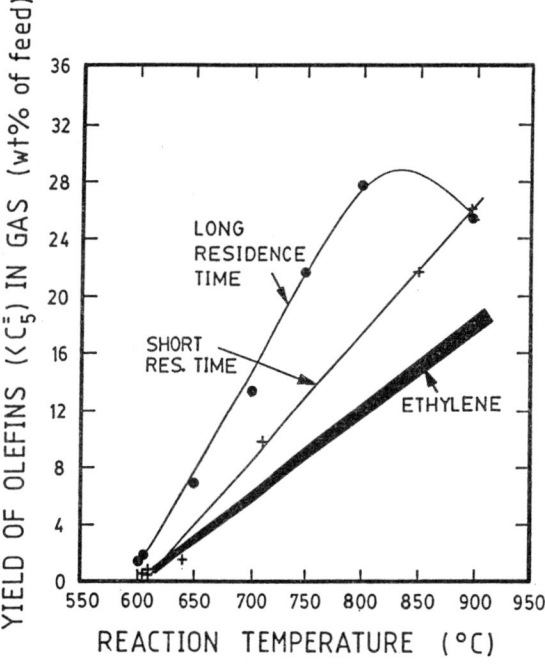

Figure 6. Plot of the yield of olefins in the total gas as a function of temperature for long (391 ± 60 ms) and short (120 ± 17 ms) residence times. The corresponding ethylene yield over the range of residence times is also shown.

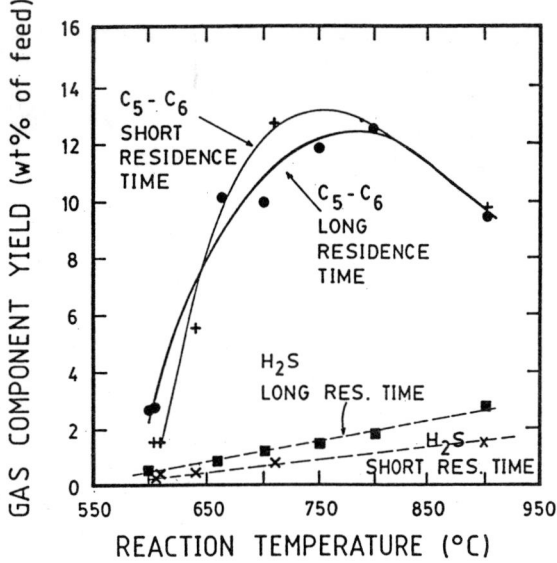

Figure 7. Plot of the $C_5$ - $C_6$ and $H_2S$ yields as a function of temperature for long (391 ± 60 ms) and short (120 ± 17 ms) residence times.

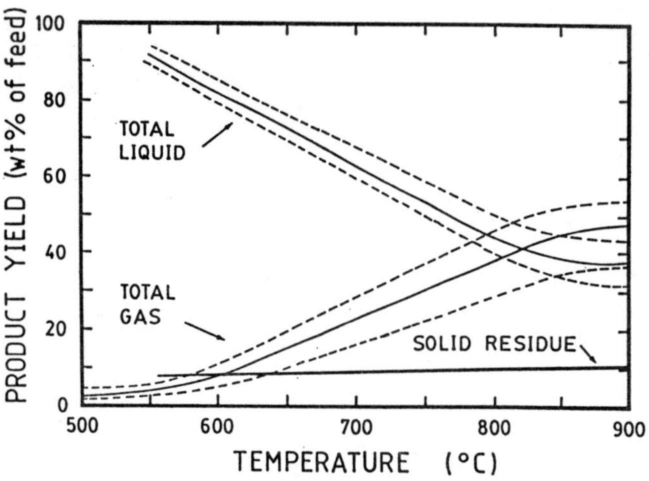

Figure 8. Plot of product yields (total gas, total liquid, solid residue) as a function of temperature. The dotted lines outlining each solid curve represent variation in the product yields due to residence time changes.

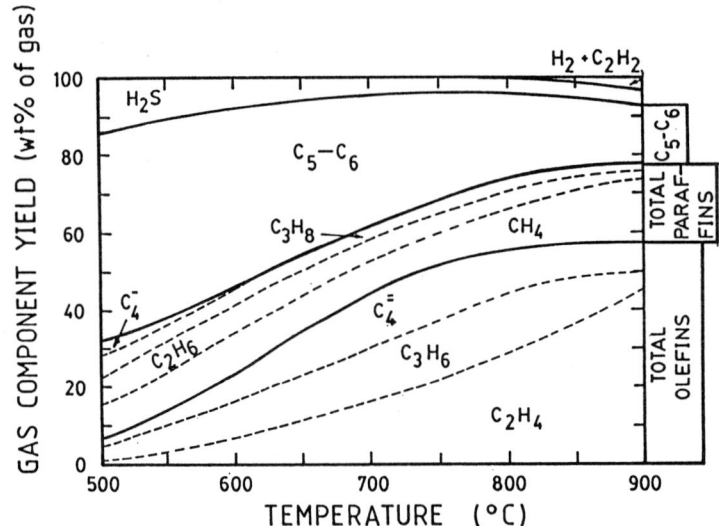

Figure 9. Variation of the gas component yields (wt% of *gas*) as a function of temperature for Cold Lake bitumen feedstock.

# PREDICTION OF SILICON POWDER ELUTRIATION IN A FLUIDIZED-BED REACTOR FOR THE SILANE DECOMPOSITION REACTION

K.Y. Li, S.H. Peng and T.C. Ho ■ Chemical Engineering Department, Lamar University, Beaumont, TX 77710

A fluidized-bed model, based on the bubble-cloud-emulsion mass transfer mechanism and a modified bubble assemblage concept, was proposed to simulate the silane decomposition reaction in a fluidized-bed CVD reactor. The model assumed that the silane homogeneous nuclear decomposition reaction occurred only in the bubble phase and the silane surface decomposition occurred in both the cloud and the emulsion phases. The effects of silane inlet concentration, gas velocity, temperature, and distributor on the silicon fine powder elutriation were examined. The results indicated that a lower gas flow rate, a lower temperature and a more perforated distributor gave a lower silicon elutriation rate. The simulation results were compared and found in agreement with experimental data available in the literature.

A cost-effective process to produce low-cost solar-cell grade silicon is the key to successful utilization of solar energy by photovoltaic conversion (Yaws, et al. 1981). After more than ten years of research activities funded by DOE/JPL, the development of chemical vapor deposition (CVD) of silane/hydrogen mixture in a fluidized-bed reactor (FBR) became a focal point for research. The CVD-FBR process appears to have great promise as an economical means of producing solar- grade silicon (Lutwack, 1986).

In the existing silane/hydrogen process developed by Union Carbide, the Siemens-type CVD reactor is used (Iya, 1986; Taylor, 1987) to deposit silicon on a hot rod of pure silicon. In the Siemens-type CVD reactor, a significant amount of fine silicon powders is formed from a homogeneous decomposition of silane. This fine powder of silicon could not be used in current crystal growers due to its low density and high risk for contamination. In order to suppress this homogeneous decomposition reaction a large excess of hydrogen gas is used in the feed. This makes the throughput of the reactor low and the production cost of silicon high.

The rough criteria for homogeneous nucleation and heterogeneous decomposition has been reported from a silane pyrolysis in a fixed-bed reactor (Iya, 1982). Generally speaking, a lower reaction temperature and lower silicon concentration in a silane/hydrogen mixture favor heterogeneous decomposition.

The fundamental mechanism of chemical vapor deposition of silane/hydrogen in a fluidized-bed reactor has been studied at a JPL-California (Hsu, 1986). The mechanisms involved were modeled as a six-path process: (1) heterogeneous deposition; (2) Homogeneous decomposition; (3) coalescence; (4) Coagulation; (5) Scavenging; and (6) chemical vapor deposition growth on fines. The heterogeneous deposition and the scavenging of the fine powders serve as the major paths for the silicon growth (Hsu et al., 1987).

An ideal backmixed reactor (CSTR) model and a fluidized-bed bubbling reactor (FBBR) model have been developed for silane pyrolysis (Dudukovic et al. 1986). After comparing with JPL experimental results, the CSTR model predicts lower fine powders formation while the FBBR model overpredicts the formation of fines.

In this study a modified bubble assemblage fluidized-bed model is used to simulate the CVD of silane/hydrogen in a FBR. With the consideration of bubble-cloud-emulsion mass transfer mechanism and the assumption that silane homogeneous nucleation occurs only in the bubble phase and the silane surface deposition occurs both in cloud and emulsion phases, the fines elutriation will be predicted. The effect of gas velocity, silane composition, temperature and gas distributor on the elutriation will be discussed. The predicted results will be compared with the JPL experimental results.

## SILANE DECOMPOSITION REACTIONS

As pointed out by Hsu and his co-workers (1987) the most important paths contributing to the surface growth of silicon are heterogeneous deposition and scavenging of fines due to the fluidization of the big particles. Therefore it is reasonable to assume that the silicon deposition reaction on the particle surface occurs only in the emulsion and cloud phases. While the homogeneous decomposition occurs mainly in the bubble phase.

Based on the above assumption, the physical configuration of a bubble and its environment is illustrated in Figure 1. As can be seen, the silane nucleation proceeds only in the void bubble phase and the surface reaction proceeds in both the cloud and emulsion phases.

During the nucleation reaction, fine powder is formed as tiny particles. Based on the JPL's experimental work (Hsu et al., 1984), the largest effluent fine particle size is less than 10 $\mu$m. Compared to the smallest packed particle size of 150 $\mu$m, it is reasonable to assume that the fine particles from nucleation are suspended in the void bubble phase and elutriated by the gas stream.

The reaction steps for the silane decomposition in fluidized-bed reactor can then be written as,

$$SiH_4 \xrightarrow{K_f} (Si) + 2H_2 \quad (1)$$

$$SiH_4 \xrightarrow{K_s} [Si] + 2H_2 \quad (2)$$

where $(Si)$ is the silicon fine particle and $[Si]$ is the silicon deposition on the silicon particle surface.

The reaction rate for reaction steps (1) and (2) can be expressed as,

$$r_f(\text{silane}) = -K_f C(SiH_4), = -\frac{1}{2}r_f(H_2) \quad (3)$$

and $$r_s(\text{silane}) = -K_s A C(SiH_4), = -\frac{1}{2}r_s(H_2) \quad (4)$$

where $K_f$ and $K_s$ are homogeneous and surface decomposition rate constants, respectively.

They can be computed from following equations (Hogness et al., 1936; Iya et al., 1982):

$$K_f = 2.0 \times 10^{13} e^{-51700/RT}, \quad 1/\text{sec} \quad (5)$$

and $$K_s = 5.14 \times 10^9 e^{-38800/RT}, \quad \text{cm/sec} \quad (6)$$

## FLUIDIZED-BED MODEL

The fluidized-bed assemblage model has been proposed by Kato and Wen (Wen and Fan 1975) and has been used to describe the hydrochlorination of silicon in a fluidized-bed reactor (Li et al. 1988). The detailed assumptions and modifications were described in the above two references. Therefore only a brief description of this model will be given below.

The bubble assemblage model simulates the bed by several consecutive compartments. The height of each compartment is determined by the bubble size in that compartment. To obtain the concentration profile the model is followed by finite difference computation from the bottom to the top. The advantages of this model are that no adjusted parameter is needed, more physical details are considered and thus can be considered as useful to scale up a laboratory fluidized-bed reactor.

With the consideration of the chemical reactions mentioned in the above section, and the bubble-cloud-emulsion mass transfer mechanism, the system equations to describe the silane/hydrogen pyrolysis in a fluidized-bed reactor may be written in the following finite difference forms

$$US(C^i_{b,n-1} - C^i_{b,n}) = F^i_{bc} V_b (C^i_{b,n} - C^i_{c,n}) + (V_b - V_c) r^i_f \quad (7)$$

$$F^i_{bc} V_b (C^i_{b,n} - C^i_{c,n}) = F^i_{c,e} V_b (C^i_{c,n} - C^i_{e,n}) + V_c r^i_s \quad (8)$$

$$F^i_{ce} V_c (C^i_{c,n} - C^i_{e,n}) = V_e r^i_s \quad (9)$$

where n is the number of compartments, i represents the component, i.e., silane or hydrogen, and b, c and e represent bubble, cloud and emulsion phases, respectively.

The volume of cloud, $V_c$, and the total volume of the bubble phase, $V_b$, can be computed from

$$V_{c,n} = \frac{N_b \Pi (\Delta h_n)^3}{6} \left( \frac{3U_{mf}/e_{mf}}{U_b - U_{mf}/e_{mf}} \right), \quad (10)$$

and $$V_{b,n} = \frac{N_b \Pi (\Delta h_n)^3}{6} \left( \frac{U_b + 2U_{mf}/e_{mf}}{U_b - U_{mf}/e_{mf}} \right), \quad (11)$$

respectively. Then the volume of the emulsion phase can be obtained as

$$V_{c,n} = S\Delta h_n - V_{b,n}. \quad (12)$$

The gas interchange coefficient between bubble and cloud and emulsion is calculated by,

$$(F_{bc})_b = 4.5 \left( \frac{U_{mf}}{d_b} \right) + 5.85 \frac{D^{\frac{1}{2}} g^{\frac{1}{4}}}{d_b^{1\frac{1}{4}}} \quad (13)$$

$$(F_{ce})_b = 6.78 \left( \frac{e_{mf} D_e U_b}{d_b^3} \right)^{\frac{1}{2}} \quad (14)$$

In the above equation, $\Delta h_n$ is the height of the n-th compartment and is calculated as,

$$\Delta h_n = 2D_o \frac{(2+m)^{n-1}}{(2-m)^n}, \quad (15)$$

where $m = 1.4\rho_s d_p (U/U_{mf})$, and $D_o$ is the initial bubble diameter which may be calculated from

$$D_o = \left( \frac{6G}{\Pi} \right)^{0.4} /g^{0.2}, \quad (16)$$

where $G = (U - U_{mf})/N_o$ and $N_o$ is the specific hole number of the gas distributor. The bubble size at height h, in a fluidized bed with $N_o$ holes per unit area on the gas distributor can be approximated as

$$D_b = 1.4\rho_s d_p (U/U_{mf}) h + D_o. \quad (17)$$

The $U_{mf}$ is the minimum fluidization velocity and is calculated from

$$\frac{1.75}{\phi_s e_{mf}^3} \left( \frac{d_p U_{mf} \rho_g}{\mu} \right)^2 + \left( \frac{150(1-e_{mf})}{\phi_s^2 e_{mf}^3} \right) \left( \frac{d_p U_{mf} \rho_g}{\mu} \right)$$
$$= \frac{d_p^3 \rho_g (\rho_s - \rho_g) g}{\mu^2}. \quad (18)$$

The number of bubbles in the n-th compartment is obtained from,

$$N_b = \frac{6S(L - L_{mf})}{\Delta h_n^2 L \Pi}, \quad (19)$$

and the bubble velocity, $U_b$, is computed from

$$U_b = 0.711(gd_b)^{\frac{1}{2}} \quad (20)$$

## COMPUTATION PROCEDURES

To obtain the concentration profile and silicon elutriation rate of silane decomposition, the following parameters should be given first:

1. Reactor pressure = 1.345 atm
2. Reactor temperature, from 650°C to 800°C
3. Reactor diameter, 2"
4. Bed height 6"
5. The specific hole number of the distributor, $N_o$, = 2 to 40 1/cm²
6. $U/U_{mf}$ = 2—6
7. Silane mole fraction at the inlet flow = 0.05~ 0.8
8. Particle size, $d_p$ = 0.02 cm
9. Homogeneous decomposition rate constant
10. Heterogeneous decomposition rate constant

The thermodynamic data of silane and hydrogen were obtained from literature (Reid et al. 1977; Yaws et al., 1981). The binary diffusivity of hydrogen and silane was calculated from Wilke and Lee's equation (Reid et al., 1977).

By knowing the inlet compositions and all of the parameters in (n- 1)-th compartment, System Equations (7) through (9) can be solved by back substitution to obtain the concentration profile for silane and hydrogen in each phase as,

$$C_{e,n}^{silane} = \frac{F_{ce}V_b}{F_{ce}V_b + V_e A K_s} C_{c,n}^{silane} = G C_{c,n}^{silane} \quad (21)$$

$$C_{c,n}^{silane} = \frac{1}{1 + \frac{AV_e K_s}{F_{bc}V_b} + \frac{F_{ce}}{F_{bc}(F_{ce}V_b + V_e A K_s)}} C_{b,n}^{silane}$$
$$= H C_{b,n}^{silane} \quad (22)$$

$$C_{b,n}^{silane} = \frac{US}{(US + V_e A K_s G H + V_c A K_s H + V_b K_f)} C_{b,n-1}^{silane} \quad (23)$$

$$C_{e,n}^{H_2} = C_{c,n}^{H_2} + 2V_e A K_s C_{e,n}^{silane} \quad (24)$$

$$C_{c,n}^{H_2} = C_{b,n}^{H_2} + \frac{2V_e A K_s C_{e,n}^{silane} + 2V_c A K_s C_{b,n}^{silane}}{F_{bc}V_b} \quad (25)$$

$$C_{b,n}^{H_2} = \frac{2V_e A K_s C_{e,n}^{silane} + 2V_c A K_s C_{c,n}^{silane} + 2V_b K_f C_{b,n}^{silane}}{US}$$
$$+ C_{b,n-1}^{H_2} \quad (26)$$

The nucleation rate is calculated by,

$$Nucl = \sum_n \{V_b K_f C_b^{silane}\}_n \quad (27)$$

and the total reaction rate can be calculated by,

$$React = \sum_n \{V_b K_f C_b^{silane} + V_c A K_s C_c^{silane} + V_e A K_s C_e^{silane}\}_n \quad (28)$$

The composition and concentration of the n-th compartment becomes the inlet of the (n+1) compartment. The properties are recalculated based on the new composition.

The calculation procedure is repeated until the bed height, $h = \sum \Delta h_n$, reaches $(2L - L_{mf})$, where $L$ is the bed height and $L_{mf}$ is the bed height at $U_{mf}$. The silicon elutriation percentage is thus obtained by,

$$Elu\% = \frac{Nucl}{React} \times 100 \quad (29)$$

## CALCULATION RESULTS

The effect of temperature on the elutriation are shown on Figure 2 for the silane mole fraction of 0.2, $N_o$=10, $d_p$=200 μm and the fluidization velocity, $U/U_{mf}$, from 2 to 6. It can be seen that as temperature increases the elutriation also increases. The increasing rate is higher at higher temperature and/or at higher fluidization velocity.

The effect of silane concentration may be seen clearly from Figure 3. At the temperature of 800°C, the elutriation decreases as the mole fraction of silane increases. However, at lower temperature this silane concentration effect is not significant. This suggests the optimum throughput for the silane decomposition in the fluidized-bed reactor may be at lower temperature and higher silane concentrations.

The effect of distributor may be seen from a plot of percent of elutriation to the number of holes of the distributor. As shown in Figure 4, when the number of holes of the distributor increases the percent of elutriation decreases and approaches to a small number for different temperatures. This suggests the smaller bubble formed the better the avoidance of fines in the silicon production. Usually several layers of stainless steel screen supported by a perforated plate is used as the gas distributor. The number of holes created by this distributor may be much more than that without a screen. Therefore this type of gas distributor with screen layers should be better than the one without screens.

## APPLICATION AND DISCUSSION

The silane decomposition in a fluidized bed has been reported by Hsu et al. (1984). The experimental data obtained from the 2" inner diameter of fluidized-bed reactor will be used in this study for comparison. The packed silicon particle size was from 150 to 300 μm. The bed height was 6 inches. The distributor was constructed by using a stainless steel plate perforated with multiple 1/32 inch holes. Either a 200 mesh stainless steel screen or none was put on the perforated plate. The actual specific number of holes is thus unknown.

The experimental data collected in a temperature range from 650 to 700°C with silane inlet compositions from 5 to 50 mole % and at a superficial velocity of $U/U_{mf}=6$ are shown in Figure 5 which are compared with the theoretical predictions which are based on the value of $N_o=20$. It can be seen that the experimental data fall into the range of predictions except for the low silane mole fraction.

Figure 6 shows the comparison of experimental data and predicted results at 700°C, the silane mole fraction of 0.65 and the gas velocity of 3 and 6 $U/U_{mf}$. As can be seen, the experimental data lies between the predictions with distributors having about 4 to 8 holes per square centimeter. This is equivalent to about 120 holes for the 2" fluidized-bed reactor.

Figure 7 shows the comparison of experimental data and predicted results with temperatures from 650 to 800°C at a gas velocity of 6 $U/U_{mf}$. It can be seen that at lower temperatures, the predictions agree with the experimental results. However, the predicted results are much higher than the experimental data at the temperature of 800°C either for $N_o=10$ or $N=20/cm^2$.

## CONCLUSION

A modified bubble assemblage model has been used successfully to simulate the silane decomposition in a fluidized-bed reactor. The model shows clearly that lower gas flow rate, lower temperature and smaller perforated holes on the gas distributor give a lower silicon elutriation percent. The effect of silane mole fraction on the elutriation percent is insignificant at a reaction temperature lower than 700°C. The simulated resulted are in agreement with experimental data available in the literature.

## ACKNOWLEDGMENT

Support for this study was provided through a Lamar University- Beaumont Organized Research Grant.

## NOMENCLATURE

$A$ Specific particle surface area, $cm^2/cm^3$
$C_b^i$ Concentration of component, i, in bubble phase, g-mol/cc
$C_c^i$ Concentration of component, i, in cloud phase, g-mol/cc
$C_e^i$ Conconcentration of component, i, in emulsion phase, g-mol/cc
$d_b$ Single bubble diameter, cm
$d_p$ Particle size, cm
$D$ Diffusivity of binary gas mixture, $cm^2/sec$
$D_e$ Effective diffusivity, $cm^2/sec$
$D_o$ Initial bubble diameter, cm
$e_{mf}$ Void fraction at minimum fluidization
$F_{bc}$ Gas interchange coefficient between bubble and cloud phase, 1/sec
$F_{ce}$ Gas interchange coefficient between cloud and emulsion phase, 1/sec
$g$ Gravitational acceleration, $cm/sec^2$
$h$ Current bed height, cm
$\Delta h_n$ Compartment height on n-th compartment, cm
$K_f$ Nucleation rate constant, 1/sec
$K_s$ Surface reaction rate constant, 1/sec
$L$ Total bed height, cm
$N_b$ Number of bubbles in the compartment
$N_o$ Specific number of holes, $1/cm^2$
$S$ cross-sectional area of reactor, $cm^2$
$U$ Superficial velocity of gas, cm/sec
$U_b$ Bubble rising velocity, cm/sec
$U_{mf}$ Minimum fluidization velocity, cm/sec
$V_b$ Total volume of bubble phase, $cm^3$
$V_c$ Total volume of cloud phase, $cm^3$
$V_e$ Total volume of emulsion phase, $cm^3$
$\mu$ Viscosity of gas mixture, cps
$\rho_g$ Density of gas phase, $g/cm^3$
$\rho_s$ Density of solid particle, $g/cm^3$
$\phi_s$ Geometric factor of particle

## LITERATURE CITED

1. Dudukovic, M., "Fluidized-Bed Reactor Modeling for Production of Silicon by Silane Pyrolysis," Proceeding of the Flat- Plate Solar Array Project Workshop on Low-Cost Polysilicon for Terrestrial Photovoltaic Solar-Cell Applications, DOE/JPL-1012-122 (1986) p. 167.
2. Hogness, T. R., T. L. Wilson, W. C. Johnson, J. Am. Ch. Soc., Vol 58, 108 (1936).
3. Hsu, G., R. Hogle, N. Rohatgi, and A. Morrison, J. Electrochem. Soc., Vol 131, 660 (1984).
4. Hsu, G., "Fluidized-Bed Development at Jet Propulsion Laboratory," Proceeding of the Flat-Plate Solar Array Project Workshop on Low-Cost Polysilicon for Terrestrial Photovoltaic Solar- Cell Applications, DOE/JPL-1012-122 (1986) p. 147.
5. Hsu, G. N. Rohatgi, and J. Houseman, AIChE Journal, Vol 33 (5), 784 (1987).
6. Iya, S. K., R. N. Flagella, F. S. Dipaolo, J. Electrochem. Soc., Vol 129, 1531 (1982).
7. Iya, S., "Development of the Silane Process for the Production of Low-Cost Polysilicon," Proceeding of the Flat-Plate Solar Array Project Workshop on Low-Cost Polysilicon for Terrestrial Photovoltaic Solar-Cell Applications, DOE/JPL-1012-122 (1986) p. 135.
8. Li, K. Y., S. H. Peng, and T. C. Ho, "Hydrochlorination of Silicon in A Fluidized-Bed Reactor," Fluidization Engineering, Fundamentals and Application-AIChE Symposium Series 262, Vol 84, p. 114 (1988)
9. Lutwack, R. "Silicon Material Task of the DOE/FSA Project," Proceeding of the Flat-Plate Solar Array Project Workshop on Low-Cost Polysilicon for Terrestrial Photovoltaic Solar-Cell Applications, DOE/JPL-1012-122 (1986) p. 3.
10. Reid, R. C., J. M. Prausnitz, and T. K. Sherwood, The Properties of Gases and Liquids, 3rd ed., McGraw-Hill Book Co., New York, 1977.
11. Wen, C. Y. and L. T. Fan, "Models for Flow Systems and Chemical Reactors," Marcel Dekker, Inc., New York, 1975.
12. Yaws, C., K. Y. Li and S. M. Chou, "Economics of Polysilicon Processes," Proceeding of the Flat-Plate Solar Array Project Workshop on Low-Cost Polysilicon for Terrestrial Photovoltaic Solar-Cell Applications, DOE/JPL-1012-122 (1986) p. 79.

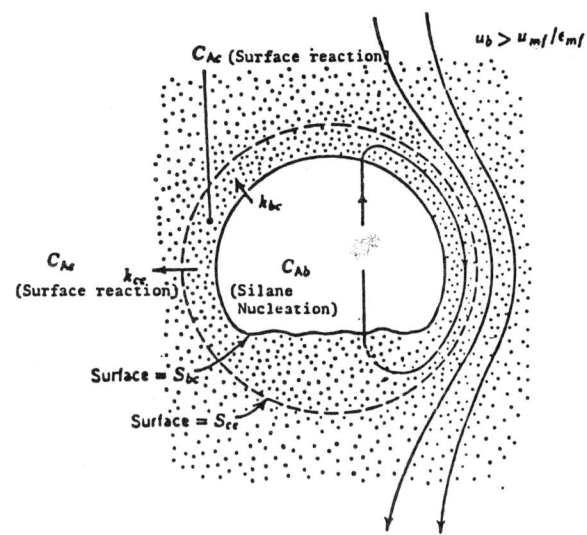

FIGURE 1. MODEL OF SILANE DECOMPOSITION

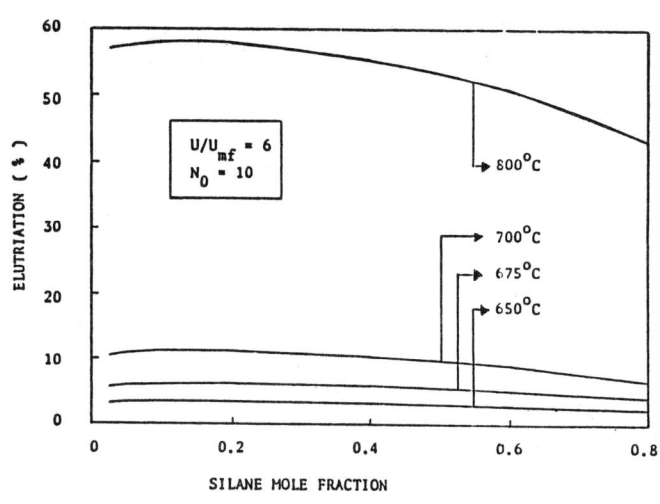

FIGURE 3. EFFECTS OF SILANE MOLE FRACTION ON ELUTRIATION

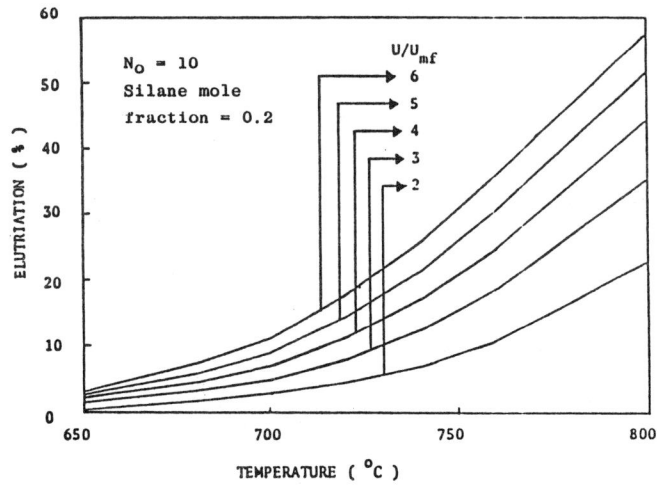

FIGURE 2. TEMPERATURE EFFECTS ON ELUTRIATION

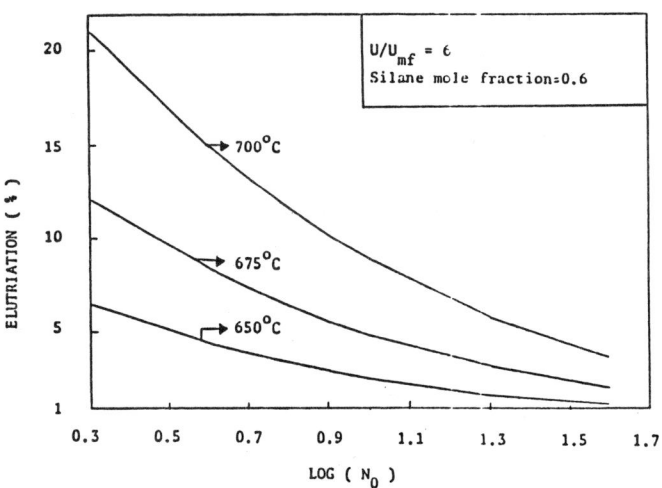

FIGURE 4. EFFECTS OF DISTRIBUTOR ON ELUTRIATION

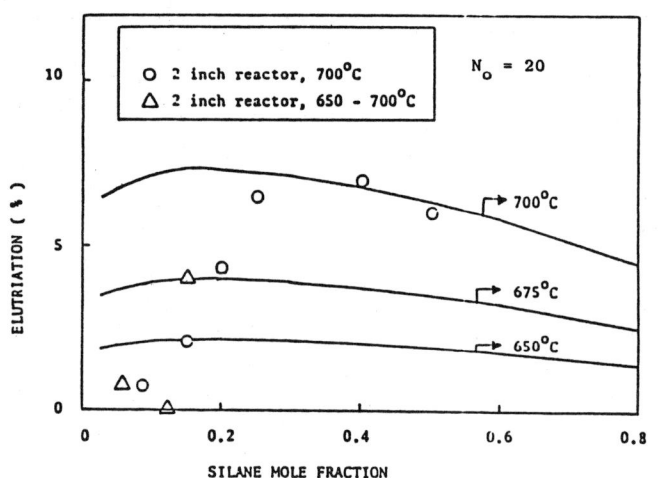

FIGURE 5. COMPARISON OF EXPERIMENTAL AND CALCULATED RESULTS AT $U/U_{mf} = 6$.

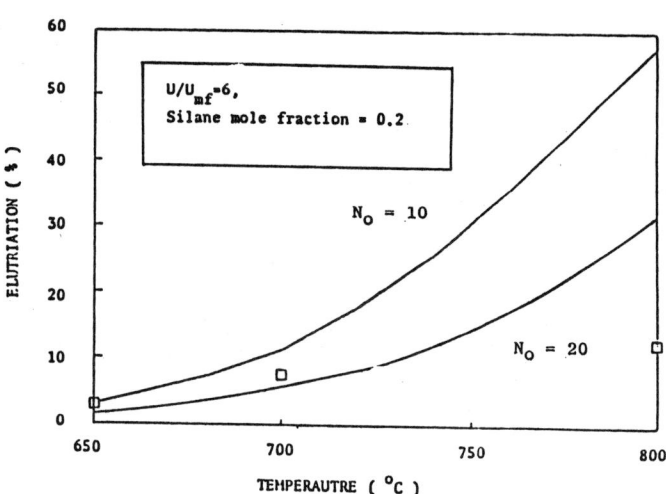

FIGURE 7. COMPARISON OF EXPERIMENTAL AND CALCULATED RESULTS AT DIFFERENT TEMPERATURES.

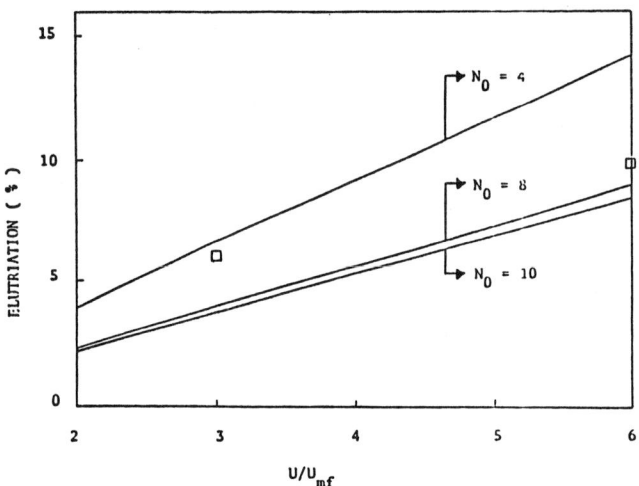

FIGURE 6. COMPARISON OF EXPERIMENTAL AND CALCULATED RESULTS AT TEMPERATURE OF 700°C AND SILANE MOLE FRACTION OF 0.65.

# THE EFFECT OF SYSTEM PARAMETERS ON FINES GENERATION IN FLUIDIZED LIMESTONE/COAL-CHAR MIXTURES

C. Sishtla, I. Chan, J. Findlay and T.M. Knowlton ■ Institute of Gas Technology, Chicago, IL 60632

Tests were conducted in a high-temperature, 6-inch-diameter test unit and a high-pressure, 6-inch-diameter unit to determine the effect of gas density on the fines generation rate of −20+35 mesh mixtures of coal chars and limestone. Both bituminous and lignite coal chars were tested. Limestone concentrations of 20, 35, and 50 weight percent were used in the investigation. Data was obtained using a multi-hole gas distributor over a temperature range of 70° to 1400 °F and a pressure range of 5 to 350 psig.

The fines generation rate of the coal chars in the fluidized bed was found to decrease with system temperature and increase with system pressure. The fines generation rate in the bed was found to be due primarily to the grid gas jets. Limestone addition to the char beds was found to reduce the rate of generation of the char fines. A successful correlation of the fines generation rate constant for both chars was obtained as a function of the kinetic energy of the grid gas jets and the Hardgrove Grindability Index (HGI).

## INTRODUCTION

Gasification processes are now emulating Fluidized Bed Combustion processes in using limestone to remove sulfur from the system. The mixing of limestone and coal in the fluidized bed can significantly change the operating characteristics of the system compared to operation with coal alone. Therefore, tests were conducted with fluidized-bed mixtures of limestone and coal char, to determine the effect of limestone concentration on fines generation in fluidized beds. The effects of gas velocity, distributor type, system temperature, system pressure, char type, and chemical reaction on fines generation in these beds were also investigated.

## EQUIPMENT, MATERIALS, AND PROCEDURE

The fines generation tests were conducted in two test units. The bulk of the tests were conducted in the 6-inch-diameter fluidized-bed test unit shown in Figure 1. This unit was only capable of low-pressure operation, but could be operated at high temperatures. The fluidization column was equipped with a non-weeping, bubble-cap-type of gas distributor (Figure 2). Six, equally-spaced bubble caps were located on a circle

---

Institute of Gas Technology, Chicago, Illinois

of 2-inch radius from the center of the column. Three, 0.0938-inch-diameter holes angled downward at 30 degrees from the horizontal were drilled in each cap.

The column was enclosed by an external electric resistance furnace that could heat the system to 1800°F. Nitrogen was used as the fluidizing gas in all tests. In the high-temperature tests, the nitrogen was passed through a preheater to reduce the load on the electric furnace. The gas passed through the preheater and the fluidization column, and then through a 1.5-inch-diameter cyclone. The gas was then exhausted to atmosphere.

The second test unit utilized was capable of operation at high pressures — up to 650 psig. A schematic drawing of this facility is shown in Figure 3. The fluidization column of the high-pressure test unit was also 6 inches in diameter. The gas distributor of this unit was exactly the same as the distributor used in the high-temperature unit. High-pressure nitrogen was passed through an orifice meter and a gas-flow control valve before entering the column. After passing through the bed, the gas passed through a cyclone and a backpressure control valve before being exhausted to atmosphere.

For chemical-reaction tests, solids were continuously added to the fluidized bed from a feed hopper located above the bed. The solids feed rate was controlled with an L-valve located immediately below the feed hopper. The solids flowed out of the bed through an overflow standpipe and were collected in a receiver located below the fluidization column.

Coke breeze (a subbituminous coal char), lignite char, and limestone were the three materials used in the investigation. All three materials were screened into size fractions of -20+30 mesh and -30+35 mesh. The actual test material was generated by mixing equal weights of the two size fractions. The blending of the two narrow size fractions allowed the same particle size distribution to be used in each test, thus minimizing sample variability. The size range of the materials was, therefore, 841 x 500 microns with a surface mean volume average diameter of 621 microns. The particle densities of the coke breeze and lignite chars were 99 and 71 $lb/ft^3$, respectively. The particle density of the limestone material was 161 $lb/ft^3$.

Three limestone concentrations were used in the testing: 20, 35, and 50 percent by weight. The complete fluidization velocities for limestone/coke breeze and the limestone/lignite char mixtures were found to be insensitive to limestone concentration within the accuracy of the determination. The complete fluidization velocities for the limestone/coke breeze and the limestone/lignite char mixtures (for all limestone concentrations) were 1.15 ft/s and 1.08 ft/s, respectively.

In each fines generation test series, the test material was fluidized at a constant superficial gas velocity until the fines generation rate became constant with time. Measurements of the fines generation rate were made at several time intervals in order to determine the fines generation rate as a function of time. For the low-pressure tests, the sequence of time intervals was: two, 1/2-hour periods followed by one, 1-hour period and then two, 2-hour periods followed by two, 4-hour periods. This sequence gave a total test time of 14 hours. For the high-pressure tests, the time intervals were substantially shorter, as was the total test time.

At the end of each time interval, the test char was removed from the column, riffled, and a particle-size analysis obtained of a representative bed sample. This sample (plus the other bed material) was returned to the test column. The fines separated from the outlet gas by the cyclone were collected, weighed, and analyzed, but were not returned to the bed. The amount of fines produced in each test was defined to be the weight percent of the char finer than 35 mesh obtained both from the bed and the char collected in the cyclone.

In all tests, both limestone and coal char fines were produced. The amount of limestone in a particular test sample was determined by reacting the limestone with HCl and measuring the amount of $CO_2$ evolved.

RESULTS AND DISCUSSION

Tests were conducted in fluidized beds of -20+35 mesh limestone/coal char mixtures to determine the effects of gas velocity, system temperature, system pressure, char type, limestone concentration, and chemical reaction on the fines generation rate. All tests (except the chemical reaction tests which were continuous-flow tests) were conducted in the batch mode using a fluidized-bed height of 12 inches in the 6-inch-diameter columns. The bubble-cap-type of distributor with the gas jets pointing downward was used because it was non-weeping, and because downward-pointing gas jets are known to cause higher rates of attrition than upward-pointing gas jets (Zenz and Kelleher[1]), thus minimizing the time necessary to produce measurable quantities of fines.

Effect of Gas Velocity

Six test mixtures were used in the investigation (20, 35, and 50 weight percent mixtures of limestone and coke breeze, and 20, 35, and 50 weight percent mixtures of limestone and lignite char). For each mixture, fines-generation tests were conducted at three superficial fluidizing gas velocities. For the limestone/coke breeze mixtures, the experimental velocities were 1.5, 2.0, and 2.25 ft/s. The superficial gas velocities used for the limestone/lignite char test mixtures were 1.25, 1.5, and 2.0 ft/s. The velocities were selected so that the lowest velocity was above the complete fluidization velocity of the mixture.

The effect of gas velocity on the rate of fines generation in the bed is shown in Figure 4 for a -20+35 mesh mixture of 35 weight percent limestone/65 weight percent lignite char. In this figure, the cumulative weight percent of fines (defined as the amount of solids generated smaller than 35 mesh) produced is plotted for each superficial fluidizing-gas velocity tested as a function of the time spent in the fluidized bed. The amount of fines produced increased significantly with the superficial fluidizing-gas velocity due to the increased kinetic energy of the gas jets from the bubble caps at the grid, and the increased momentum of the particles colliding with each other in the bed.

For all three curves shown in Figure 4, the initial rate of fines production was high, but then leveled off to a relatively constant rate after about three hours of exposure time in the bed. This is typical behavior and has also been reported by Vaux and Keairns[2], Vaux and Schruben[3], and Kono[4]. This behavior suggests that the initial fines generation rate was due to the breakage of sharp edges on the char particle which would lead to a rounding of the particles, such as observed by Blinchev et al.[5].

### Effect of Distributor Type

Tests to determine the relative amount of attrition occurring because of 1) turbulence in the bulk of the fluidized bed, and 2) the gas jets at the grid were also conducted. These tests were carried out in a test column equipped with a porous-plate distributor with a -20+35 mesh mixture of 35 weight percent limestone/65 weight percent lignite char. With the porous-plate distributor, high-velocity grid jets are not formed. Therefore, the char attrition measured in the porous-plate tests was primarily the attrition due to the turbulence in the bed. Typical results of these tests are shown in Figure 5, along with the results from the test with the same material in the column fitted with the bubble-cap distributor. The fines-generation curves shown in Figure 5 were both obtained at a constant superficial fluidizing gas velocity of 2.0 ft/s. The results show that the amount of fines generated in the tests without the grid-gas jets was substantially less (approximately an order of magnitude) than the amount of fines generated in the tests with the grid-gas jets. This was observed for both the limestone/coke breeze and the limestone/lignite char mixtures at all limestone concentrations. Thus, the majority of the coal-char fines were produced by the grid-gas jets, and not by the turbulence in the fluidized bed. Blinichev et al.[5] and Chen et al.[6] have also observed that most of the fines in a fluidized bed were produced at the distributor.

### Effect of System Temperature

Experiments were also conducted to determine the effect of system temperature on the fines generation rate in limestone/coke breeze and limestone/lignite char test mixtures. Tests were conducted at system temperatures of 65°, 800°, and 1400°F. Typical results showing the effect of temperature on the fines generation rate of char are shown in Figure 6 for a -20+35 mesh mixture of 35 weight percent limestone/65 weight percent coke breeze at a constant fluidizing-gas velocity of 2.0 ft/s. In this figure, the cumulative weight percent of coke breeze fines produced at temperatures of 65°, 800°, and 1400°F is plotted as a function of time the solids were in the fluidized bed. The figure shows that the amount of fines generated decreases significantly with increasing system temperature.

The reason for the decrease in the fines generation rate with temperature is that the major source of fines generation in the batch fines generation tests is the grid-gas jets, and that fines generation at the grid is proportional to the kinetic energy of the gas jets ($\rho_g U_j^2$). Because the rate of fines generation is increased, $\rho_g$ decreases and the rate of fines generation decreases. Kono[4] also found that the fines generation rate decreased with increasing temperature because of the temperature effect on gas density.

### Effect of System Pressure

Tests were also conducted with the limestone/coal char mixtures at system pressures of 60 and 175 psia. Typical results of the tests are shown in Figure 7. In this figure, the cumulative weight percent of coke breeze fines produced in the tests is plotted as a function of time for system pressures of 15, 60, and 175 psia for comparison. Both tests were conducted at a constant superficial fluidizing-gas velocity of 1.5 ft/s with a 35 weight percent limestone/65 weight percent coke breeze test mixture. The results show that the amount of coke breeze fines that were generated was significantly greater at the higher pressure. Approximately 60 percent of the original coke breeze material was converted to fines in only about 1.5 hours at 175 psia. In contrast, at 15 psia, only about 10 percent of the original coke breeze material was converted to fines after a total time in the bed of 14 hours.

The significant effect of pressure on the fines generation rate of coke breeze was expected. As explained above, the fines generation rate at the grid is proportional to the kinetic energy of the grid-gas jets. Because the kinetic energy is proportional to $\rho_g$, when the system pressure is increased, the fines generation rate is increased.

### Effect of Limestone Concentration

The effect of limestone concentration on fines generation in limestone/coal char mixtures was investigated for each char. Typical results are shown in Figure 8 for limestone/coke breeze mixtures. In this figure, the cumulative weight percent of coke breeze fines produced in the tests was plotted as a function of time for limestone concentrations of 0, 20, 35, and 50 weight percent at a superficial fluidizing gas velocity of 2 ft/s. The percentage of coke breeze fines generated was calculated as a percentage of the total weight of coke breeze charged to the test column (not the total weight of the limestone/coke breeze mixture). Figure 8 shows that the percentage of coke breeze fines generated decreased with increasing limestone concentration.

It was expected that the coke breeze fines generation rate would decrease when limestone was added to the bed. This is because in a homogeneous limestone/coke breeze mixture, the amount of coke breeze contacting the gas jets in the grid region of the fluidized bed would only be a fraction of the coke breeze contacting the grid gas jets in a fluidized bed of 100% coke breeze material operating at the same conditions. However, the percentage of coke breeze fines generated with limestone present was found to be less than expected (i.e., it was less than the percentage of coke breeze fines generated with the 100% coke breeze material multiplied by the percentage of coke breeze fines in the test material with limestone present). For example, after 14 hours, approximately 57% of the 100% coke breeze material was converted to fines (-35 mesh material) at a superficial fluidizing-gas velocity of 2 ft/s, while only about 25% of the 65% coke breeze material was converted to fines (Figure 8). This is only about 44% of the fines generated as a percentage of initial feed material instead of the expected 65%. An explanation for the discrepancy was that the heavier limestone segregated and was present in higher concentrations than the average bed limestone concentration in the grid region. The segregation scenario, in conjunction with a theoretical analysis of how the presence of limestone in the bed lowers the fines generation rate, provides a way to predict the experimental results.

As discussed above, the char fines generation rate produced by solids collisions in the bulk of the fluidized bed is negligible compared to the fines generation due to the grid-gas jets. Near the grid, a solid particle is theoretically accelerated to a high velocity by the jet, and then breaks the bed particle with which it collides, as well as breaking up itself. With 100% coke breeze as the bed material, two coke breeze particles are broken with each collision. With limestone present in the bed, two coke breeze particles are broken only when a coke breeze particle collides with another coke breeze particle. When a limestone particle collides with a coke breeze particle, only one coke breeze particle is broken.

Using this explanation of how the limestone affects char fines generation, the amount of coke breeze fines generated as a percentage of initial coke breeze feed when limestone is present, can be predicted using the results of the 100% coke breeze fines generation test. For a generalized limestone/coke breeze mixture of X weight fraction coke breeze and (1-X) weight fraction limestone, the percentge of coke breeze fines produced relative to the 100% coke breeze mixture at the same conditions and cumulative

$$X \cdot X \cdot F_{100} + ([1-X]X \cdot F_{100})/2 = Y \quad (1)$$

where

$X$ = the weight fraction of coke breeze in the mixture at the grid

$1-X$ = the weight fraction of limestone in the mixture at the grid

$F_{100}$ = the percentage of -35 mesh coke breeze fines generated with 100% coke breeze material (basis: weight percent of initial coke breeze material charged to the bed)

$Y$ = the percentage of -35 mesh fines generated in the limestone/coke breeze mixture (basis: weight percent of initial coke breeze material charged to the bed)

The first term in Equation (1) results from the fact that only X weight fraction of the material in the bed is coke breeze and can collide with X weight fraction of the material above it to produce breakage of two coke breeze particles. This product is then multiplied by the percentage of -35 mesh coke breeze fines produced in the 100% coke breeze fines generation test to give the percentage of coke breeze fines produced by the coke-coke collisions in the limestone/coke breeze mixture. Similarly, the second term calculates the percentage of -35 mesh coke breeze fines produced by limestone-coke breeze collisions. However, because only one coke breeze particle is fractured in these collisions, the term is divided by 2.

In Equation (1), the weight fraction of coke breeze and limestone are those actually present at the grid. If segregation occurs, the weight fractions of coke breeze and limestone will be different than their average weight fractions in the mixture.

Tests were conducted to determine the extent of limestone segregation in fluidized beds of 20, 35 and 50 weight percent limestone. In all three tests, the bed was operated at a superficial fluidizing-gas velocity of 2 ft/s. In each test, a measurement was made of the percentage of limestone in the bottom 2 inches and the top 2 inches of a 9.5-inch-high fluidized bed fitted with a porous-plate distributor. The measurement was made by abruptly shutting off the fluidizing gas to the bed and vacuuming out the two test sections of the bed to be analyzed. The results of the segregation tests are shown in Table 1. Values of the coke breeze fines percentages calculated from Equation (1) and the measured limestone concentrations at the grid, predicted the measured fines generation values within 20%.

## The Effect of Chemical Reaction

Tests were also conducted to determine the effect of chemical reaction on the fines generation rate of lignite char in fluidized beds. To enable a valid comparison of fines generation, with and without chemical reaction, tests were first conducted at 1400°F without chemical reaction to determine the "base-case" rate of fines generation in the 6-inch-diameter fluidization test unit as a function of superficial fluidizing gas velocity. Three tests were then conducted with chemical reaction (i.e., with 10% oxygen in the fluidizing gas) at the same fluidizing gas velocities as used in the first three tests. The results of the tests were then compared to determine the effect of chemical reaction on the fines generation rate of lignite char in fluidized beds of limestone/lignite char mixtures.

The conditions and results of the tests without chemical reaction are summarized in Table 2, and the conditions and results of the tests with chemical reaction are shown in Table 3. The rate of fines generation was expressed as pounds of lignite char fines generated per pound of lignite char fed to the unit. From the results shown in Tables 2 and 3, this parameter was greater in the tests with chemical reaction than it was in the tests without chemical reaction. This can be seen more clearly in Figure 9, in which the pounds of lignite char fines produced per pound of lignite char fed to the reactor is plotted as a function of the kinetic energy of the grid gas jets. At the lowest grid gas kinetic energy (corresponding to a superficial fluidizing gas velocity of 1.25 ft/s) the rate of fines generation with chemical reaction was approximately equal to the rate without chemical reaction. At the highest grid jet kinetic energy (corresponding to a superficial fluidizing gas velocity of 2.0 ft/s) the rate of fines generation with chemical reaction was approximately 1.5 times the rate without chemical reaction.

If the two lines in Figure 9 are extended, they intersect at a grid jet kinetic energy of approximately 1200 lb/ft-s$^2$ instead of zero. This kinetic energy value corresponds to a superficial fluidizing gas velocity of about 1.2 ft/s. This is approximately the velocity (1.08 ft/s) that was found to be necesary to completely fluidize the lignite char/limestone mixture. Therefore, practically, one cannot realistically talk of signifiant fines generation at velocities below this value — it is a packed bed at lower velocities. The fines generation rate should be very low in a packed bed because the solids will not be able to be picked up in the gas jet and collide against the other particles in the bed.

The addition of oxygen to the bed is the only difference between the two sets of tests shown in Tables 2 and 3. The increased rate of fines generation in the second group of tests is caused by the reaction between the oxygen, and the carbon in the particles. Therefore, an equation was proposed to predict the fines generation rate with chemical reaction if the fines generation rate without chemical reaction is known. The equation is

$$G_W = G_{WO} + G_{WO}(K[M - M_O]) \quad (2)$$

where

$G_W$ = pounds of lignite char fines produced per pound of lignite char fed with chemical reaction

$G_{WO}$ = pounds of lignite char fines produced per pound of lignite char fed without chemical reaction

$K$ = constant

$M$ = moles of oxygen per pound of lignite char fed

$M_O$ = moles of oxygen per pound of lignite char fed at the complete fluidization velocity (1.08 ft/s for this mixture)

This equation relates the fines produced with chemical reaction to the fines produced without chemical reaction and the moles of oxygen per pound of lignite char fed to the system. Rearranging the equation to the form

$$G_W - G_{WO} = G_{WO}(K[M - M_O]) \quad (3)$$

relates the difference in the fines generation rate with and without chemical reaction to the product of the fines generation rate without chemical reaction and the moles of oxygen used per pound of lignite char fed to the unit. This equation is plotted in Figure 10 using the data generated in the tests described above. This equation should go through zero, because at $M_O$ (corresponding to a superficial fluidizing gas velocity of 1.08 ft/s) the fines generated with and without chemical reaction are essentially the same (Figure 9). As can be seen, a relatively good fit of the data was obtained with this equation.

### Effect of Char Type

The effect of char type was determined by comparing the fines generation rate for limestone/lignite char test mixtures with the fines generation rate for limestone/coke breeze test mixtures. The comparison is shown in Figure 11, where the cumulative weight percent of fines for both coke breeze and lignite char is plotted as a function of time at a superficial fluidizing-gas velocity of 1.5 ft/s. Significantly more fines were produced from the limestone/lignite char test mixtures than from the limestone/coke breeze test mixtures at the same fluidizing gas velocity.

Both chars were subjected to Moh Hardness and Hardgrove grindability tests to try to relate the differences in the fines generation rates between the two chars to some physical property of the chars. Both chars had a Moh Hardness value of 4.5. The Moh Hardness is a measure of the abrasiveness of a material, and is chracterized by a scale of from 1 to 10 — where talc has a Moh Hardness of 1 and diamond a Moh Hardness of 10. The Hardgrove grindability test results in an index called the Hardgrove Grindability Index (HGI), which is a measure of how easy it is to grind a material. The larger the HGI, the easier it is to grind the material. Coke breeze had an HGI of 38, and lignite char an HGI of 64. Clearly, the Moh Hardness test was not useful in explaining the difference in the fines generation rates between the two chars. However, the Hardgrove grindability tests indicated that the lignite char was more likely to break under impact, and abrasion, which was what was observed experimentally. The HGI of the chars was then used to develop a correlation to predict a fines generation rate constant for the two chars.

### Fines Generation Rate Constant

The total coal char fines generation rate constant ($k_t$) in a fluidized bed can be defined by the following equation:

$$\frac{dW}{dt} = -k_t W \quad (4)$$

where

$\frac{dW}{dt}$ = rate at which coal char fines are generated in the fluidized bed, g/h

$W$ = total weight of coal char in the bed, g

$k_t$ = total coal char fines generation rate constant, $h^{-1}$

Integration of Equation (4) leads to the following expression:

$$\ln[(100-F)/100] = -k_t t \qquad (5)$$

where

F = cumulative weight percent of coal char converted to fines
= $([W_o-W]/W_o) \times 100$, %

$W_o$ = initial weight of coal char in the bed, g

t = exposure time of coal char in the bed, h

A plot of ln (100-F) versus t yields a straight line with a slope equal to $-k_t$.

The total coal char fines generation rate constant can be partitioned into a coal char fines generation rate constant due to the grid-gas jets ($k_j$) plus a coal char fines generation rate constant ($k_b$) due to turbulence in the bulk of the fluidized bed, Sishtla et al.[10], or —

$$k_t = k_j + k_b \qquad (6)$$

by subtracting the amount of coal char fines generated in the tests without grid-gas jets (i.e., those generated by the turbulence in the bed) from the total amount of fines generated in the bed, the amount of fines generated by the grid jets can be calculated. The fines generation rate constant due to the grid jets could then be calculated from the rate of fines generated by the grid jets. Sishtla et al.[10] correlated the values for the fines generation rate constant due to the gas jets at the grid with both $\rho_g U_j$ (the "momentum" of the grid-gas jets) and $\rho_g U_j^2$ (the "kinetic energy" of the grid-gas jets). Sishtla et al.[10] found that $k_j$ correlated better with $\rho_g U_j^2$ than $\rho_g U_j$.

The finding that the fines generation rate is proportional to $\rho_g U_j^2$ agrees with the results of Zenz and Kelleher[1] and broadly with those of Blinichev et al.[5]. In contrast, Kono[4] found that his measured fines generation rates were proportional to the cube of the superficial gas velocity ($U^3$).

As discussed above, the physical property which best chracterized the relative tendency of the chars to attrit, was the HGI. The HGI was incorporated by Sishtla et al.[10] into the rate constant analysis to develop an overall equation for $k_j$ which would apply to both chars and which is shown below —

$$k_j = 2.8 \times 10^{-12} (\rho_g U_j^2)^{1.56} (HGI)^{2.15} \qquad (7)$$

A similar type of analysis was performed with the limestone/coke breeze and the limestone/lignite char fines generation rate data. The fines generation rate constants $k_j$, $k_b$, and $k_t$ were calculated for both coke breeze and lignite char at the various test conditions. However, one additional variable (the concentration of coal char in the bed) was found to be necessary to correlate the fines generation rate constant for the grid gas jets ($k_j$) in limestone/coal char mixtures. Incorporating this variable into the analysis resulted in the following expression for $k_j$ for limestone/coal char mixtures.

$$k_j = 2.32 \times 10^{-13} (\rho_g U_j^2)^{1.53} (HGI)^{2.88} (X)^{1.0} \qquad (8)$$

This correlation predicted all of the fines generation data (except the data from the high-pressure tests) to within ±30%. Equation (8) predicts that the fines generation rate at the grid should increase with increasing system pressure because of the gas density increase. Although the fines generation rate was found to increase significantly with system pressure, the increase was not as large as Equation (8) would predict.

This could have been due to the fact that the fines which were carried out of the test unit and collected by the cyclone were not added back into the bed for the next fines generation test. Therefore, the amount of material in the bed was continually decreasing. The fines generation rate in the bed is directly proportional to the amount of material in the bed. In the high-pressure tests, a significant amount of fines were collected in the cyclone. Therefore, the fines generation rate was significantly affected.

ACKNOWLEDGEMENT

The authors would like to express their appreciation to the U.S. Department of Energy which funded the work. The work was performed under Contract No. DE-AC21-85MC22061.

NOTATION

F = cumulative weight percent of solids converted to fines, %

$F_{100}$ = percentage of coal char fines

generated with 100% coal char material, %

$G_w$ = pounds of coal char fines produced per pound of coal char fed (with chemical reaction), dimensionless

$G_{wo}$ = pounds of coal char fines produced per pound of coal char fed (without chemical reaction), dimensionless

K = constant

$k_b$ = fines generation rate constant due to bed turbulence, $h^{-1}$

$k_j$ = fines generation rate constant due to grid-gas jets, $h^{-1}$

$k_t$ = total fines generation rate constant, $h^{-1}$

M = moles of oxygen fed per pound of coal char fed, moles/lb

$M_o$ = moles of oxygen fed per pound of coal char fed at the complete fluidization velocity, moles/lb

t = exposure time of solids in the bed, h

U = superficial fluidizing-gas velocity, ft/s

$U_j$ = grid-gas jet velocity, ft/s

$U_{cf}$ = complete fluidization velocity, ft/s

$U_{mf}$ = minimum fluidization velocity, ft/s

W = weight of solids in the bed, g

$W_o$ = initial weight of solids in the bed, g

X = weight fraction of coal char in the bed near the grid, dimensionless

Y = percentage of fines generated with 100% coal char, %

$\rho_g$ = gas density, $lb/ft^3$

## LITERATURE CITED

1. Zenz, F. A. and Kelleher, E. G., J. of Powder & Bulk Solids Technology, 4, 13 (1980).

2. Vaux, W. G. and Keairns, D. L., in Fluidization, ed. J. R. Grace and Matsen, J. M., Plenum, New York, 437 (1980).

3. Vaux, W. G. and Schruben, J., A.I.Ch.E. Symp. Ser., 222 (79), 97 (1983).

4. Kono, H., A.I.Ch.E. Symp. Ser., 205 (77), 96 (1981).

5. Blinichev, V. N., Strel'tsov, V. V. and Lebedeva, E. S., Internat'l Chem. Eng. 8 (4), 615 (1968).

6. Chen, T. P., Sishtla, C., Punwani, D. and Arastoopour, H., in Fluidization, ed. J. R. Grace and J. M. Matsen, Plenum, New York (1980).

7. Merrick, D. and Highley, J., A.I.Ch.E. Symp. Ser., 137 (70), 366 (1974).

8. Vaux, W. G., Proc. Am. Power Conf., 40, 793 (1978).

9. Ray, Y. C., Jiang, T. S. and Wen, C. Y., Powder Technology, 49, 193 (1987).

10. Sishtla, C., Findlay, J., Chan, I and Knowlton, T. M. Paper to be presented at the Fifth International Fluidization Conference, Banff, Canada. May (1988).

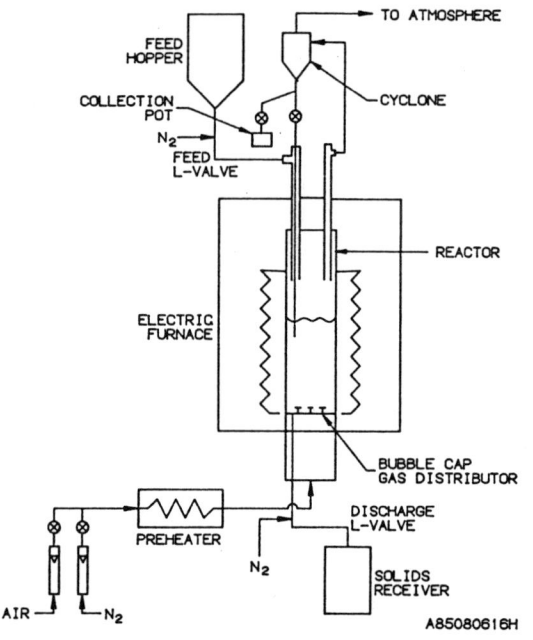

Figure 1. Schematic drawing of high-temeprature, low-pressure fines generation test unit.

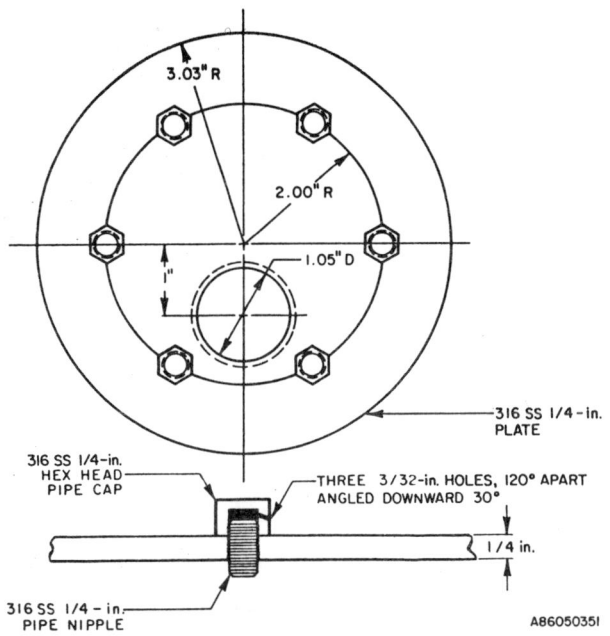

Figure 2. Distributor plate configuration.

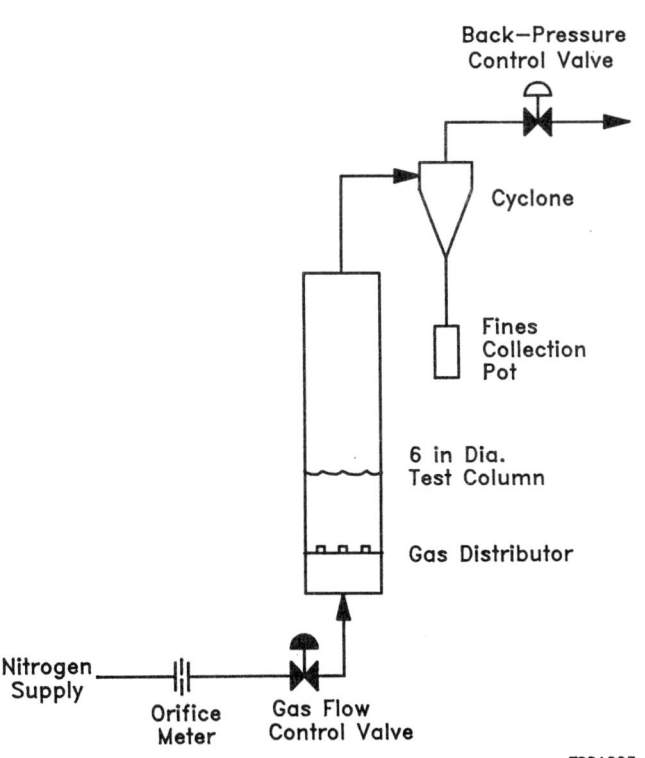

Figure 3. Schematic drawing of high-pressure test unit.

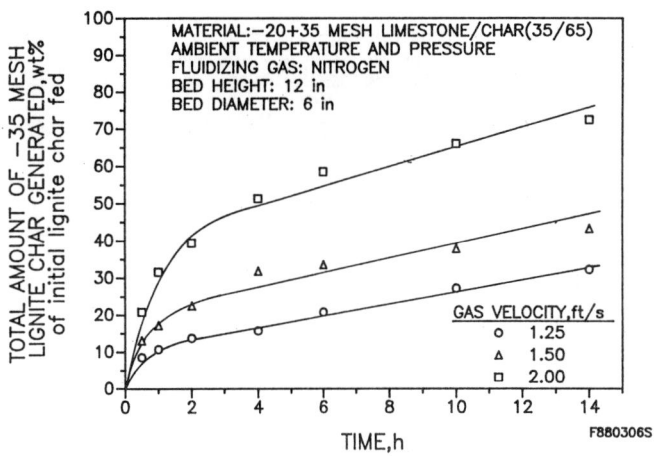

Figure 4. The effect of gas velocity on fines generation rate.

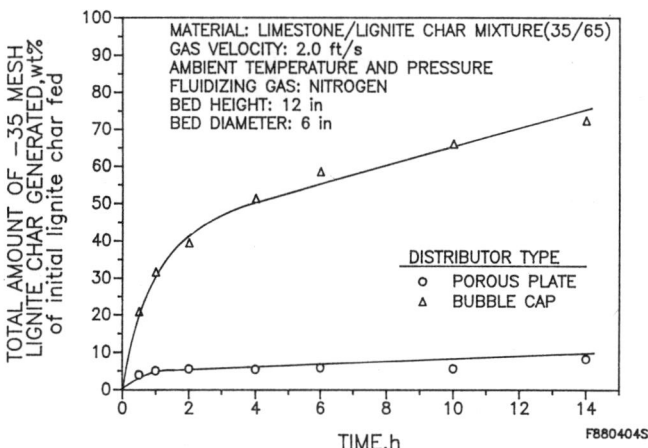

Figure 5. The effect of distributror type on fines generation rate.

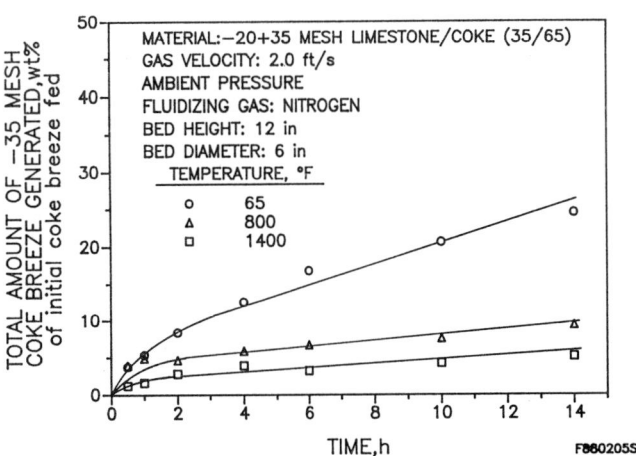

Figure 6. The effect of system temperature on fines generation rate.

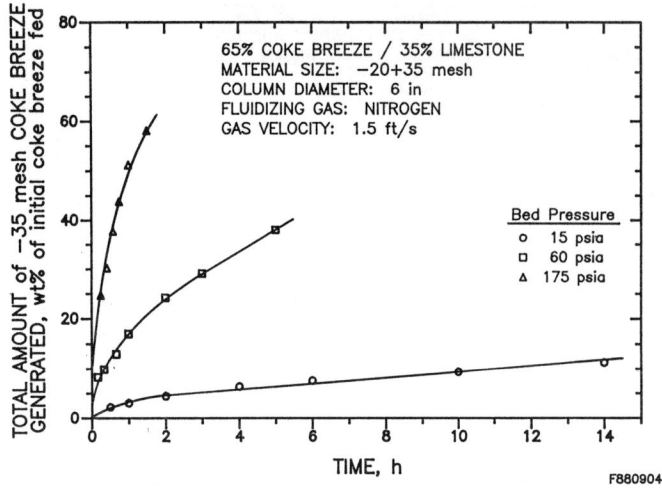

Figure 7. The effect of system pressure on fines generation rate.

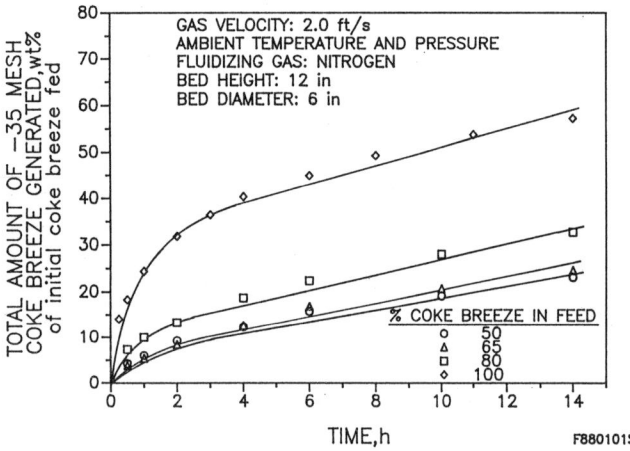

Figure 8. The effect of limestone concentration on fines generation rate.

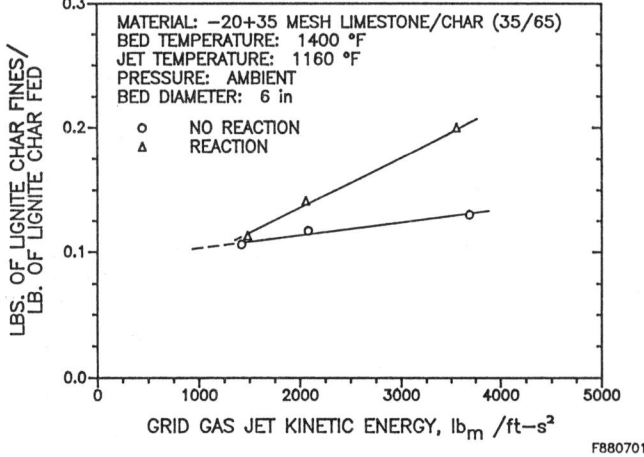

Figure 9. Comparison of lignite char fines generation rates with and without chemical reaction.

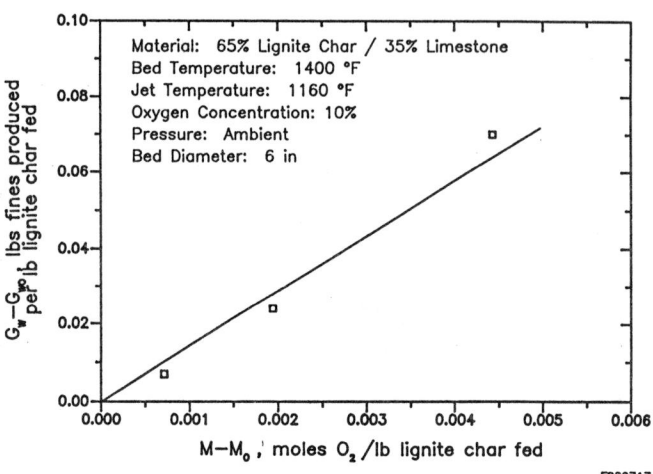

Figure 10. The difference between lignite char fines generated with and without reaction as a function of the moles of oxygen fed to the reactor.

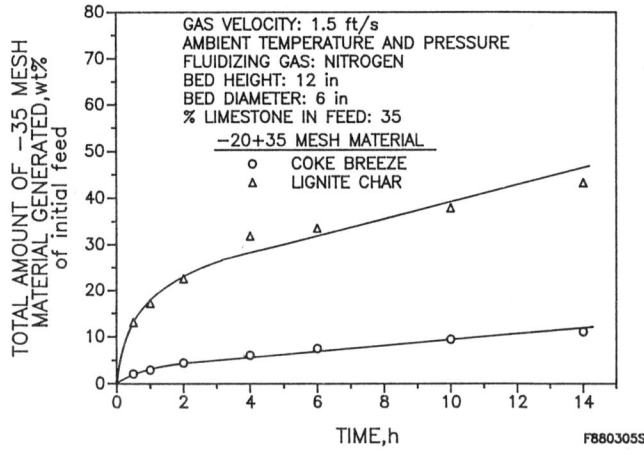

Figure 11. Relative fines generation rate for lignite char and coke breeze.

Table 1. FLUIDIZED-BED SEGREGATION TEST RESULTS WITH LIMESTONE/COKE BREEZE MIXTURES (Gas Velocity: 2 ft/s).

| Limestone in Mixture, % | Limestone Percentage, % | |
|---|---|---|
| | Top 2 Inches | Bottom 2 Inches |
| 20 | 12.9 | 25.8 |
| 35 | 35.7 | 42.1 |
| 50 | 48.7 | 56.2 |

Table 2. RESULTS AND CONDITIONS OF CONTINUOUS-FEED LIMESTONE/LIGNITE CHAR TESTS WITHOUT CHEMICAL REACTION

| | HTC-11 | HTC-9 | HTC-10 |
|---|---|---|---|
| Bed Temperature, °F | 1400 | 1400 | 1400 |
| Fluidizing Velocity, ft/s | 1.25 | 1.50 | 2.00 |
| Average Mixture Feed Rate, lb/h | 14.6 | 14.4 | 14.3 |
| Average Residence Time, h | 0.50 | 0.49 | 0.48 |
| Total Rate of Lignite Char Fines Generation, lb/h | 1.011 | 1.096 | 1.207 |
| Pounds of Lignite Char Fines Produced Per Pound Fed | 0.106 | 0.117 | 0.130 |

Table 3. RESULTS AND CONDITIONS OF CONTINUOUS-FEED LIMESTONE/LIGNITE CHAR TESTS WITH CHEMICAL REACTION

| | HTC-14 | HTC-13 | HTC-12 |
|---|---|---|---|
| Fluidizing Gas Oxygen Concentration, % | 10 | 10 | 10 |
| Bed Temperature, °F | 1400 | 1400 | 1400 |
| Fluidizing Velocity, ft/s | 1.25 | 1.50 | 2.00 |
| Average Mixture Feed Rate, lb/h | 15.1 | 15.3 | 15.1 |
| Average Residence Time, h | 0.43 | 0.47 | 0.45 |
| Total Rate of Lignite Char Fines Generation, lb/h | 1.113 | 1.407 | 1.962 |
| Pounds of Lignite Char Fines Produced Per Pound Fed | 0.113 | 0.141 | 0.200 |

# LABORATORY TESTING OF A FLUIDIZED-BED DRY-SCRUBBING PROCESS FOR THE REMOVAL OF ACIDIC GASES FROM A SIMULATED INCINERATOR FLUE GAS*

W.M. Bradshaw, R.P. Krishnan and J.M. Young ■ Oak Ridge National Laboratory, Oak Ridge, TN
G.B. Mohrman ■ U.S. Army Toxic and Hazardous Materials Agency, Aberdeen Proving Ground, MD

A series of bench-scale tests was conducted to evaluate a dry, fluidized-bed, scrubbing process for removing acidic gases from incinerator flue gas. The acidic gases studied were sulfur dioxide, hydrogen chloride, and phosphorus pentoxide. These gases were found to react readily with lime in a bubbling bed operating at 540 to 650 °C (1000 to 1200 °F). Superficial gas velocity, bed temperature, bed depth, sorbent type, and sorbent conversion strongly affected the degree of acidic gas removal. Sorbent utilization was inhibited by the occlusion of the particle surface by reaction products.

The Oak Ridge National Laboratory (ORNL) has been investigating dry fluidized-bed scrubbing of acidic compounds from incinerator flue gas under the sponsorship of the U.S. Army Toxic and Hazardous Materials Agency (USATHAMA). The requirement for scrubbing acidic flue gases arises due to the U.S. Army pursuing incineration as a means of disposing of contaminated sludges, soils, and liquids. These wastes frequently contain chemical compounds that produce acidic gases when incinerated. Common acidic gases that are formed include sulfur dioxide ($SO_2$), phosphorus pentoxide ($P_2O_5$), hydrogen chloride (HCl), and nitrogen oxides ($NO_x$). In most cases, the acidic compounds must be removed from the flue gas to meet environmental standards.

This paper summarizes experimental work done at ORNL evaluating the removal of acidic gases from a simulated incinerator flue gas, using a fluidized bed of lime [CaO or $Ca(OH)_2$] operating in the bubbling-bed regime between 430 to 650°C (800 to 1200°F). The experimental program focused on determining the feasibility of dry scrubbing $P_2O_5$, HCl, and $SO_2$ in a fluidized bed both (1) as individual gases at the same process conditions, and (2) in a mixture representative of a typical incinerator flue gas. Another objective of the tests was to determine the ultimate sorbent utilization.

BACKGROUND

For this discussion, "dry scrubbing" is defined as a process that does not produce a liquid waste. Dry scrubbing has several advantages over wet scrubbing, the most significant of which is that it produces no liquid effluents that need to be treated. This aspect makes dry scrubbing highly suitable for use at Remedial Action sites that have no water treatment facilities. Another benefit of dry scrubbing is that the temperature of the off-gas is above its dew point, thus reducing plume opacity, eliminating the need for reheat, and in some cases, reducing corrosion of downstream equipment. Dry scrubbing is the preferred alternative for incinerators burning "mixed" wastes containing certain radionuclides, because scrubber water could contribute to a criticality incident ([1]).

A considerable amount of research has been conducted by the electrical power generation industry on dry flue-gas desulfurization. Dry scrubbing has been applied to hazardous waste incinerators at several

---

*Research sponsored by the U.S. Army Toxic and Hazardous Materials Agency through the U.S. Department of Energy under contract No. DE-AC05-84OR21400 with the Martin Marietta Energy Systems, Inc.

regional, noncommercial incinerators in Europe (2). Dry scrubbing has also been used at ORNL to remove hydrogen fluoride (HF) from vent-gas streams in a bench-scale, packed-bed reactor using lime at 135 to 300°C (275 to 570°F) (3). However, few data exist in the literature concerning the dry scrubbing of $P_2O_5$, either alone or with other acidic gases.

Potential sorbents include sodium compounds such as sodium carbonate, sodium bicarbonate, nahcolite, and trona; calcium compounds such as lime, limestone ($CaCO_3$), and dolomitic quicklime ($CaO \cdot MgO$); and, to a lesser degree, magnesium compounds such as magnesium oxide. Calcium-based sorbents are preferable because (1) they are ~8.5 times less expensive (on a molar equivalent basis) than the sodium-based (alkali) sorbents (2); (2) many waste products of calcium sorbents [e.g., $CaSO_4$, $Ca_3(PO_4)_2$, and $CaF_2$] are insoluble in water (this is of particular benefit if the spent sorbent is to be disposed of in a landfill); and, (3) sodium-based sorbents are generally limited to temperatures below 370°C (700°F) because of the likelihood of sintering and melting.

A typical dry fluidized-bed scrubbing system would include: a fluidized bed as the primary gas-solids contactor, a multi-cyclone for solids removal, dilution air for cooling, and a baghouse to remove particulates. In addition to removing particulates, the baghouse also provides additional residence time for acidic gas-sorbent contacting, thus improving the acid removal rate.

EXPERIMENTAL

The experiments focused on dry scrubbing $P_2O_5$, HCl, and $SO_2$, both individually and as a mixture in a bubbling bed. Military-specific wastes that the Army may incinerate will produce all three acidic gases.

The experimental program consisted of 24 tests, executed in two phases. The objectives of the first phase were:

1. determine the feasibility of removing $P_2O_5$ in a fluidized bed of alkaline material,

2. determine an "operating envelope" for the dry scrubbing of $P_2O_5$, and

3. verify that $SO_2$ can be scrubbed at the same process conditions.

The objectives of the second phase were:

1. verify that HCl can be scrubbed at the same process conditions,

2. screen candidate sorbents,

3. determine scrubber performance with a mixture of acidic gases that represents a typical flue gas composed of multiple contaminants, and

4. obtain the data necessary to evaluate the economics of the dry-scrubbing process and establish design criteria for a pilot-scale scrubber.

The test conditions studied for $P_2O_5$ and HCl scrubbing were chosen to coincide with conditions reported in the literature that are favorable for $SO_2$ removal (4). The acid gas concentrations covered the range that might result from incineration of Army wastes. The parameters studied included bed temperature, sorbent type, inlet concentration, residence time, and gas velocity. The parameter ranges are shown in Table 1.

Table 1. Test conditions

| Parameter | Range tested |
|---|---|
| Bed temperature, °C (°F) | 480-650 (900-1200) for $P_2O_5$<br>540 (1000) for HCl<br>540 (1000) for $SO_2$<br>430-540 (800-1000) for mixtures |
| Inlet concentration, ppm | <1.0-1900, $P_2O_5$<br><138-6900, HCl<br><18-11900, $SO_2$ |
| Gas velocity, cm/s | 4.8-9.5 for $P_2O_5$<br>6.8 for HCl<br>5.7-6.8 for $SO_2$<br>2.6-8.5 for mixtures |
| Bed depth, cm (in.) | 20 (8) for $P_2O_5$<br>20 (8) for $SO_2$<br>20 (8) for HCl<br>2.5-20 (1-8) for mixtures |
| Sorbent particle size, μm | 212-425 for all tests |

Several calcium-based sorbents were considered and tested, including limestone, lime, and dolomitic quicklime. Lime was used in the majority of tests because it is stable in the temperature range of interest. In addition, the 212- to 425-μm lime particles exhibited minimal channeling and slugging. Unslaked and slaked lime supplied by the Tenn-Luttrell Company in Luttrell, Tennessee, were 92.2% CaO, 7.4% Ca(OH)$_2$, and 0.5% MgO; and 87% Ca(OH)$_2$ and 13% CaCO$_3$, respectively.

A schematic diagram of the dry scrubber system is shown in Figure 1.

The fluidized bed was contained in a 2-in., Schedule 40, 316L stainless steel pipe 46 cm (18 in.) long. A 4-in., Schedule 40 pipe 76 cm (30 in.) long was provided immediately above the 2-in. section to reduce solids entrainment. The inner wall of the scrubber was plated with nickel to guard against any corrosion that would bias the experimental data.

Solid P$_2$O$_5$ was sublimed into a nitrogen stream, and the gaseous mixture entered the scrubber via a porous metal (316L stainless steel) tuyere plate designed to enhance gas distribution and support the lime. Both SO$_2$ and HCl were introduced through a port immediately below the tuyere plate. Another line carried N$_2$, CO$_2$, and O$_2$ to the same port. All gases (with the exception of P$_2$O$_5$) were controlled with in-line rotameters. The amount of P$_2$O$_5$ fed into the scrubber was a function of the sublimation chamber temperature and the nitrogen flow rate through the sublimation chamber.

The inlet and effluent gas streams were bubbled into gas-washing bottles containing either distilled water or sodium hydroxide, and the contents were subsequently analyzed to determine the acidic gas concentrations. The sample bottles were changed at regular

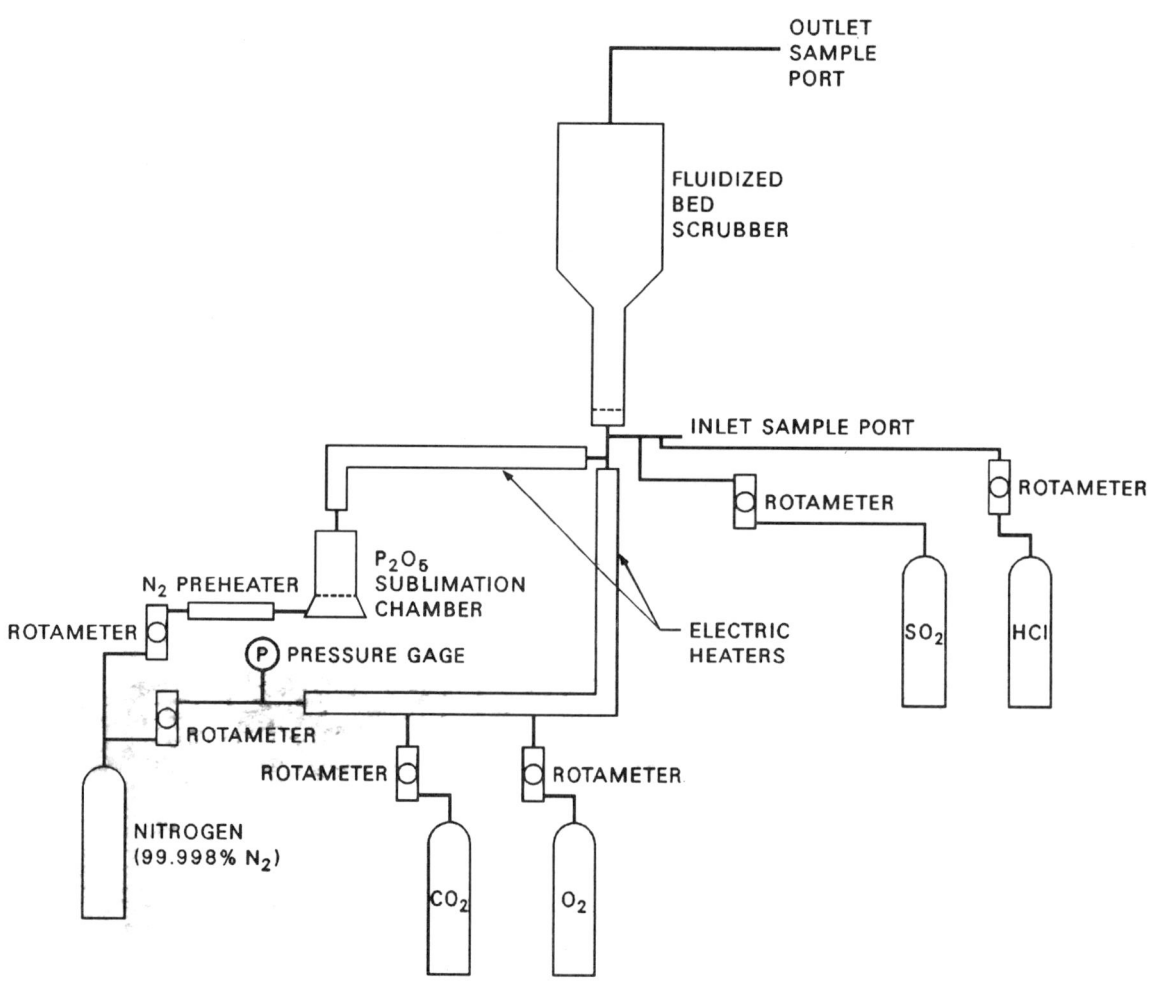

Fig. 1. Schematic diagram of experimental dry scrubber.

intervals to provide time-weighted averages of inlet and outlet concentrations.

All the tests were of 8-h duration, with the exception of two long-term tests (80 h and 112 h, respectively). The tests consisted of heating the scrubber and associated equipment and then bringing all of the flows to steady-state. The inlet gas was sampled before and after each run. Samples of the inlet and outlet gas streams were also obtained periodically throughout each run. A typical test would produce four 1-h outlet gas samples and five 30-min inlet gas samples.

The sample train was tailored to the specific gas being sampled. Phosphorus pentoxide and HCl are both soluble in water, so a single gas-washing bottle filled with distilled water was used to trap these gases when they were tested individually. Sulfur dioxide is only slightly soluble in water, but is easily scrubbed by a 1-$\underline{N}$ sodium hydroxide solution. Scrubbing all three acid gases in the same test necessitated using a 1-$\underline{N}$ sodium hydroxide solution in the gas-washing bottles because of the insolubility of $SO_2$ in water.

The sample bottles were analyzed colorimetrically for total phosphorus. Total sulfate and chloride were determined by ion chromatography (IC). Total chloride in the solids was determined by IC also. The carbonate, sulfate, and water content of the sorbent were determined by X-ray powder diffractometry.

DISCUSSION AND RESULTS

Superficial Gas Velocity

Superficial gas velocity had a pronounced effect on the HCl and $P_2O_5$ removal rate, but not on the $SO_2$ removal rate. The relationship between removal rate and gas velocity (at constant bed depth) is shown in Figure 2. Both HCl and $P_2O_5$ removal rates increased with gas velocity. The $SO_2$ removal rate decreased slightly with velocity.

The relationship between superfical gas velocity and removal rate might be explained on the basis of the rate-limiting steps involved in the single particle gas-solid reaction. The overall removal rate is a function of (1) external gas film diffusion between the bulk gas and solid particle, (2) diffusion within the particle pores, and (3) chemical reaction at the solid surface. In general, the gas film diffusion (external mass transfer) increases as gas velocity increases. The intraparticle diffusion and

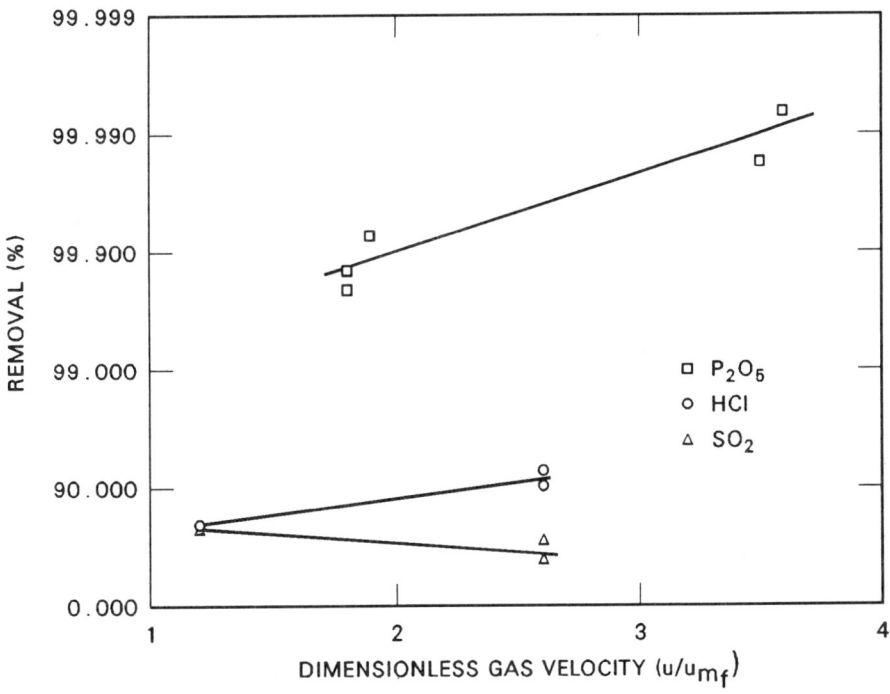

Fig. 2. Acid gas removal versus gas velocity (at constant bed depth).

reaction rates are not affected by gas velocity. It is proposed that the HCl and $P_2O_5$ removal rate is controlled by external diffusion while the $SO_2$ removal rate is controlled by intraparticle diffusion or chemical reaction. This is not unexpected because both HCl and $P_2O_5$ are much stronger acids than $SO_2$. Lower $SO_2$ removal at higher gas velocity (and constant bed depth) is probably related to decreased gas residence time. Other researchers have reported a similar relationship between $SO_2$ removal and gas residence time for a fluidized bed combustor that used calcined limestone as the sorbent (5).

### Bed Temperature

Bed temperature was also found to affect the removal rate. The removal rate for $P_2O_5$ increased slightly with temperature between 480 and 540°C, but exceeded 99% at all conditions tested. Both HCl and $SO_2$ showed a much more distinct temperature dependence. Removal rate data for fresh sorbent are shown in Table 2.

Table 2. Effect of bed temperature on removal

| Gas | Temperature (°C) | Gas velocity $(u/u_{mf})^a$ | Removal (%) |
|---|---|---|---|
| $SO_2$ | 427 | 1 | 21.0 |
| $SO_2$ | 540 | 1 | 78.4 |
| HCl | 427 | 1 | 8.8 |
| HCl | 540 | 1 | 79.9 |
| $P_2O_5$ | 480 | 2 | 99.7 |
| $P_2O_5$ | 540 | 2 | >99.8 |
| $P_2O_5$ | 650 | 2 | >99.4 |

$^a U_{mf}$ experimentally measured at temperature from pressure drop vs velocity data.

### Sorbent Type

Lime, limestone, and pulverized dolomitic quicklime (57% CaO, 40% MgO) were tested for their effectiveness in removing acid gases. As expected, uncalcined limestone was a very poor sorbent in the chosen temperature range. At 540°C and a superficial gas velocity of 5.7 cm/s ($u/u_{mf} = 2$), only 1% of the HCl and 2% of the $SO_2$ reacted with the sorbent. A single test was conducted with HCl in nitrogen using dolomitic quicklime at 540°C and a gas velocity of 6.8 cm/s ($u/u_{mf} = 2.6$). The removal rate exceeded 98%, which is comparable to results obtained with calcined lime at the same conditions. Lime was selected as the sorbent for the remaining tests.

### Residence Time

In general, higher acid gas removal was observed as the residence time increased. For example, in tests using a 2.5-cm (1-in.) lime bed (static bed depth) at minimum fluidization (2.6 cm/s) and 540°C, HCl removal averaged 79.9%. Increasing the bed depth to 10-cm (4-in.) at the same temperature and gas flow rate resulted in an average HCl removal of 95.5%.

### Inlet Gas Concentration

The acid gas removal rate was independent of inlet gas concentration at the conditions studied. However, at a given set of conditions, the removal rate was higher for a single acidic gas in an inert carrier gas than when other acidic gases were also present.

### Sorbent Utilization

Two long-term tests were conducted to determine the relationship between sorbent utilization and gas removal. The first one, run at 540°C and 3.6 times minimum fluidization, tested the removal of $P_2O_5$ from nitrogen using CaO as the sorbent. Total test duration was 80 h. The average removal exceeded 99.99%. No decrease in removal rate was observed over the range of 0 to 9% sorbent utilization.

The second long-term test (112 h) included all three acidic gases as well as $N_2$, $O_2$, and $CO_2$ in ratios representative of incinerator flue gas. The sorbent utilization at the conclusion of the second test was 71% (based on sorbent solids analysis). The acidic gas removal rate was strongly dependent on utilization. When HCl was the only acidic gas present, the removal was consistently higher than when $SO_2$ was also present. The relationship between removal efficiency and sorbent utilization (using the gas data) is shown in Figure 3.

During the 112 h of testing, 32% of the lime escaped from the bed (based on a calcium balance). The particles collected in the outlet filter were estimated to be <50 μm. For the most part, the loss was due to elutriation of fines from the bed. A 24-h blank

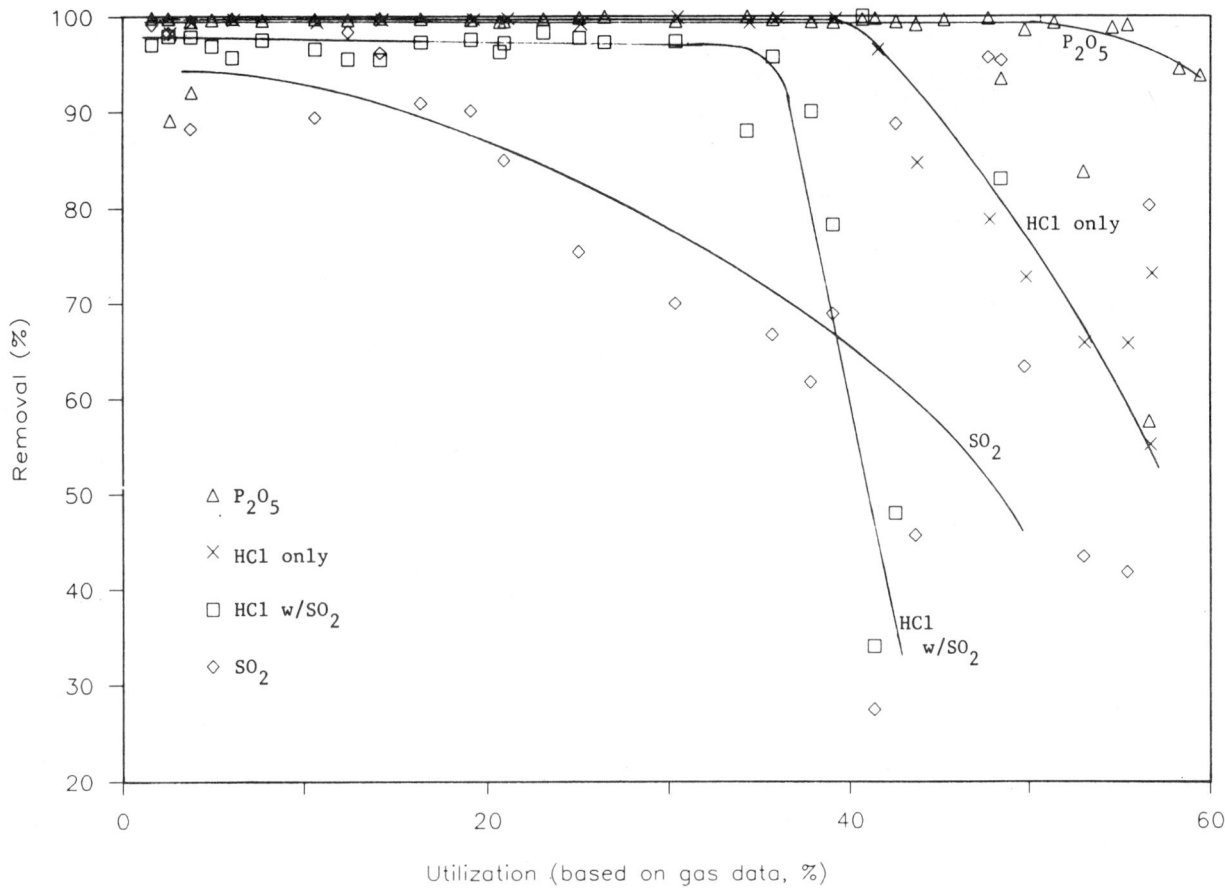

Fig. 3. Acidic gas removal efficiency versus sorbent utilization.

run was conducted with an unreacted lime test bed using nitrogen to fluidize the bed at the same flow rate and temperature. About 15% elutriation was observed for the unreacted lime.

At the conclusion of the second long-term test, the sorbent contained slightly more $CaCl_2$ and $CaSO_4$ than predicted. A material balance using the gas data indicated that 36.8 g of chloride and 30.3 g sulfate should have been captured by the sorbent. The sorbent actually contained 39.4 g chloride and 36.0 g sulfate at the end of the test. These differences fall within the measurement error for the equipment and procedures that were used in the experimental program.

As sorbent utilization increases, the pores become increasingly occluded with reaction product. This is clearly shown in the micrographs of particles used in the second long-term test. Figure 4 shows a typical lime particle (at 750X) before reacting with the sorbent. Figure 5 shows a spent particle from the second long-term test (the utilization averaged 71% for the bulk material). The micrographs strongly suggest that, as the

Fig. 4. Scanning electron micrograph of lime particle before test at 750X.

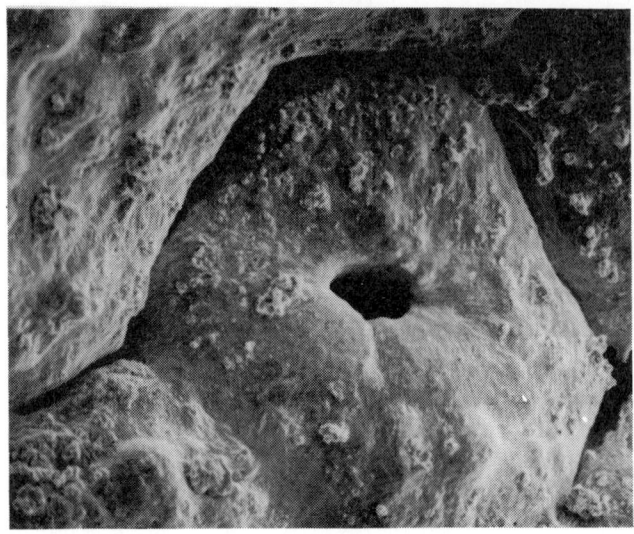

Fig. 5. Scanning electron micrograph of lime particle after test at 750X.

reaction proceeds, spent particles tend to become coated on the surface and further reaction is impeded. The BET surface area for the raw and spent lime is 3.93 $m^2/g$ and 1.87 $m^2/g$, respectively.

Surface occlusion is typically the Achilles' heel of dry scrubbing processes that use lime or limestone. Surface coating is inevitable because the volume per mole of calcium for $CaO$, $Ca_3(PO_4)_2$, $CaCl_2$, and $CaSO_4$ is 17, 33, 52, and 63 $cm^3$, respectively. The high molar volume of $CaSO_4$ makes $SO_2$ a particularly difficult gas to remove. The analytical laboratory reported that many particles were almost completely coated with $CaSO_4$; it was virtually impossible to detect any other compounds on the surface. By contrast, the surface of the sorbent used in the long-term test, in which $P_2O_5$ was the only acidic gas, showed no tendency toward plugging. A scan of the particle surface after the test showed <1% phosphorus (below detectable limits) although the bulk sorbent contained 1.8% phosphorus. It is believed that the bulk of the $Ca_3(PO_4)_2$ was formed within the pores of the lime particles and did not tend to coat the surface.

## CONCLUSIONS

Bench-scale tests support the feasibility of dry, fluidized-bed scrubbing of flue gas containing HCl, $P_2O_5$, and $SO_2$. This process should be considered for acidic gas removal, particularly when use, handling, or disposal of aqueous scrubber effluents is unsafe or abnormally expensive. Pilot-scale testing is a reasonable next step.

The acid gas removal process involves external mass transfer, intraparticle diffusion, and reaction; each of which depend on parameters such as temperature, gas velocity, residence time, and sorbent characteristics. The relationship reported here between removal rate and superfical gas velocity demonstrates that the rate-limiting step at a particular set of conditions can differ for various acidic gases. Any attempt to extrapolate to a different set of conditions must address the potential for changes in the rate-limiting step.

Maintaining accessible particle surface area is a key to maximizing sorbent utilization. Increased particle abrasion and attrition at higher gas velocity is beneficial to removal efficiency, provided the loss due to elutriation is controlled. A circulating fluidized bed (CFB) would likely be a better candidate from this standpoint. High gas velocity and extensive particle attrition in the CFB would serve to minimize particle coating and therefore increase the availability of the sorbent.

## LITERATURE CITED

1. S. Carpentier et al., *Design and Operation of Radioactive Waste Incineration Facilities*, International Atomic Energy Agency Draft Report, 134-35 (1987).

2. P. J. Kroll and P. Williamson, "Application of Dry Flue Gas Scrubbing to Hazardous Waste Incineration," *J. Air Pollut. Control Assoc.*, 36(11), 1258-63 (1986).

3. C. H. Brown and V. L. Fowler, *Removal of Hydrofluoric Acid from Gas Streams by Solid Sorbents*, ORNL/TM-9243, Martin Marietta Energy Systems, Inc., Oak Ridge Natl. Lab. (1985).

4. L.-S. Fan et al., "Limestone/Dolomitic Sulfation in a Vertical Pneumatic Transport Reactor," *Ind. Eng. Chem. Process Des. Dev.*, 23(3), 539-45 (1984).

5. G. P. Zimmerman et al., *Atmospheric Fluidized-Bed Combustion Testing of Western Kentucky Limestones*, ORNL/TM-8383, Union Carbide Nuclear Division, Oak Ridge Natl. Lab. (1982).

# DEVELOPMENT OF A CIRCULATING FLUIDIZED BED FERMENTOR: THE HYDRODYNAMIC MODEL FOR THE SYSTEM

D. Pirozzi, L. Gianfreda and G. Greco, Jr. ■ Dipartimento di Ingegneria Chimica dell' Università, Napoli, Italy
L. Massimilla ■ Dipartimento di Ingegneria Chimica dell' Università e Istituto di Ricerche sulla Combustione del CNR, Napoli, Italy

The hydrodynamic model of a liquid-operated circulating fluidized bed has been developed in view of possible applications to fermentation processes. The model consists in four blocks: fast column, hydrocyclone, recirculation column and solids recirculation control valve. Changes in the behavior of liquid-operated circulating fluidized beds as compared to gas-operated ones have been discussed.

Circulating fluidized beds (CFB) have been successfully used to enhance gas-solid contact in a variety of processes in chemical and energy industries [Basu, 1986] [Basu and Large, 1988]. Little information, however, is available in the literature as regards the application of CFB technique to liquid-solid and gas-liquid-solid systems, though an entirely new field of application could be envisaged by carrying out fermentation processes in CFB reactors with immobilized biomass supported onto solids.

Relevant feature of ordinary fluidized bed fermentors is the uniformity in bed properties that results from effective mixing in the solid phase. CFB operation further promotes solids mixing, thus increasing product throughput per unit bed cross-section. CFB fermentors could also operate with liquid streams of different nature in the fast column and in the recirculation column, should this be required by process development. Furthermore, attrition phenomena promoted by CFB flow patterns might produce a continuous abrasion of the biofilm, whose fragments may be separated in the hydrocyclone. In case of fast biomass growth, this feature could prevent fermentor clogging.

A hydrodynamic model of the CFB system is discussed in the present work. The model is similar to that developed by Arena et al. (1987) for CFB systems operated with air.

## MODELLING THE CFB SYSTEM

The system is schematically shown in Fig. 1. It consists of four blocks: the fast column, the hydrocyclone, the recirculation column and the solids control valve. The fast column is fed with a high velocity liquid stream. The recirculation column has the dual purpose of storing solids while providing the driving force for their circulation. Liquid is injected through a porous plate distributor at the bottom of the column and helps in keeping the descending solids in a state of gentle fluidization. Solids are introduced into the fast column by means of an inclined pipe. A butterfly valve keeps solids feed rate at the desired value. Liquid and solids are separated in the collecting section downstream the fast column by means of an hydrocyclone.

Model assumptions are that:

- the system is isothermal;
- flow is steady and monodimensional;
- mass and heat transfer resistances between phases are negligible;
- liquid is newtonian, with constant density and viscosity in each block;
- hydrocyclone efficiency is equal to 1;

According to Rudd and Watson (1968), the system has been divided into its component blocks (circles in Fig. 2). The mass balance over all the blocks (square in Fig. 2) has been taken into account as an additional block.

The geometrical variables of the system, namely: $D_{FC}$, $L_{FC}$, $D_{RC}$, $L_{RC}$, $D_{IP}$, $A_{FC} = \pi D_{FC}^2/4$, $A_{IP} = \pi D_{IP}^2/4$, (Fig. 1) are assigned, and therefore do not appear in the CFB information structure (Fig. 2). The same is true for the variables characterizing the hydrocyclone (e, b, $D_{CY}$, $L_{CY}$, n and $\gamma$), for the solids hold-up in the hydrocyclone ($H_{CY}$) and for that in the valve ($H_{VA}$). As shown by calculations, system sensitivity to variations in voidage $\epsilon_{FC0}$ at the entrance of the fast column is small, so that a reasonable value of 0.75 has been assumed, regardless of the values of other variables. Similarly, values of 0.5 and 0.02 have been assumed for voidages $\epsilon_{RCbed}$ of the fluidized bed contained in the recirculation column and $\epsilon_{RCfb}$ of suspension in the free-board of such column, respectively. The pressure downstream the cyclone $P_{RC1}$ is 1 bar since the system discharges into the atmosphere.

Design relationships, together with design and state variables, will be discussed in detail for each block (Tab. 1).

Fast Column. The fast column is described by the two continuity equations 1 (liquid phase) and 2 (solid phase), together with the two momentum balance equations 3 (overall mixture) and 4 (solid phase).

Variables involved in equations 1 and 2 are: voidage $\epsilon_{FC}(x)$, liquid velocity $V_L(x)$, solid velocity $V_S(x)$, liquid superficial velocity $U_L$ and mass flux of solids $G_S$. The latter two variables are constant within the fast column since both phases are incompressible.

Equation 3 is the mixture momentum balance, where $F_W$ is the pressure gradient due to friction between liquid-solid mixture and fast column wall. The latter results from two separate contributions: $F_{WL}$ for the liquid phase and $F_{WS}$ for the solids [Teo and Leung, 1984].

Equation 4 is the solid momentum balance [Soo, 1967] where $F_S$ is the drag force per unit volume of particles [Teo and Leung, 1984]. Pressure $P_{FC}(x)$ is an additional variable in equation 3, to be considered together with $\epsilon_{FC}(x)$, $V_L(x)$, $V_S(x)$, $U_L$ and $G_S$. It should be pointed out that $F_W$ and $F_S$ are not independent variables, since both depend on other variables that have been already introduced.

The entrance pressure $P_{FC0}$ is the only variable to be supplied in order to determine $V_L(x)$, $V_S(x)$, $P_{FC}(x)$, $\epsilon_{FC}(x)$ since $\epsilon_{FC}(0) = 0.75$ and, according to equations 1 and 2, $V_L(0)\epsilon_{FC}(0) = U_L$ and $V_S(0)(1-\epsilon_{FC}(0)) = G_S/\rho_S$. The corresponding profiles are obtained by solving the system of algebraic and first order differential equations (eqns. 1 to 4) as an initial value problem. Altogether, local degrees of freedom for this component block are 3, namely: 7 variables ($V_L(x)$ or $V_{L1}$, $V_S(x)$, $P_{FC}(x)$ or $P_{FC1}$, $\epsilon_{FC}(x)$ or $\epsilon_{FC1}$, $U_L$, $G_S$, and $P_{FC0}$) minus 4 relationships (eqns. 1 to 4).

**Hydrocyclone**. The pressure drop correlation by Rietema (1961) involves several variables, i. e. liquid velocity $V_{L1}$, voidage $\epsilon_{FC1}$, pressure $P_{FC1}$ at the top of the fast column, pressure $P_{RC1}$ at the top of the recirculation column. There are 3 local degrees of freedom, namely: 4 variables ($V_{L1}$, $\epsilon_{FC1}$, $P_{FC1}$ and $P_{RC1}$) minus 1 design relationship (eqn. 5).

**Recirculation Column**. The recirculation column equation provides the total pressure drop along the column by taking into account solid-liquid head in the fluidized bed and in the free-board. In this system block there is 1 local degree of freedom, namely: 2 variables ($P_{RC0}$ and the column bed height h, being $P_{RC1} = 1$) minus 1 design relationship (eqn. 6). Note that $P_{RC1}$ is treated differently in the recirculation column and in the hydrocyclone, though it is the same variable. Indeed, in the recirculation column it is a datum of the problem, whereas in the hydrocyclone it stands as a recycle variable, as explained later in the description of the information structure of Fig. 2.

**Control Valve**. The control valve equation consists in a momentum balance along the inclined pipe. Liquid and solid phases have been assumed to have the same velocity. The equation is a modification of the solid-liquid fluidized bed orifice equation by Massimilla et al. (1963). ($P_{RC0} - P_{FC0}$) is the pressure drop through the valve, $\phi$ is the valve opening, (ratio between cross sectional area available to the flow, and total cross sectional area $A_{IP}$). If unity efflux coefficient C is assumed, the local degrees of freedom are 3, namely: 4 variables ($G_S$, $P_{RC0}$, $P_{FC0}$ and $\phi$) minus 1 design relationship (eqn. 7).

**Solids Mass Balance**. Solids inventory is distributed throughout the CFB unit. Therefore, there are 2 local degrees of freedom, namely: 3 variables (H, h and $\epsilon_{FC}(x)$) minus 1 design relationship (eqn. 8).

## TABLE 1. CFB DESIGN EQUATIONS

### Fast Column

(1) $V_L(x) \ \epsilon_{FC}(x) = U_L$

(2) $V_S(x) \ (1-\epsilon_{FC}(x)) = G_S/\rho_S$

(3) $-dP_{FC}(x)/dx = (1 - \epsilon_{FC}(x)) \ \rho_S \ V_S(x) \ dV_S(x)/dx \ \epsilon_{FC}(x) \ \rho_L \ V_L(x) \ dV_L(x)/dx + \varsigma \rho_S \ (1 - \epsilon_{FC}(x)) + \rho_L \ \epsilon_{FC}(x)] \ g + F_W$
where:
$F_W = F_{WL} + F_{WS}$

(4) $F_S \ (1 - \epsilon_{FC}(x)) = (1 - \epsilon_{FC}(x)) \ \rho_S \ V_S(x) \ dV_S(x)/dx + (\rho_S - \rho_L) \ (1 - \epsilon_{FC}(x)) \ g + F_{WS}$

### Hydrocyclone

(5) $(P_{FC1} - P_{RC1}) = C_{in} f \ (Re_{in}) \ \gamma \ (b/e)^n \ (D_{CY}/L_{CY})^{0.7} \ (O/Q)^{0.8}$
where:
$f(Re_{in})$ is an empirical function of Reynolds number at the inlet of the hydrocyclone
$C_{in} = 0.5 \ [\rho_S \ (1 - \epsilon_{FC1}) + \rho_L \ \epsilon_{FC1}] \ [4 \ V_{L1} \ A_{FC} / (\pi b^2)]^2$
$Q = A_{FC} \ (U_L + G_S/\rho_S)$
$O = Q - A_{FC} \ G_S / [\rho_S \ (1 - \epsilon_{FCfb})]$

### Recirculation Column

(6) $(P_{RC0} - P_{RC1}) = [\rho_S \ (1 - \epsilon_{FCbed}) + \rho_L \ \epsilon_{FCbed}] \ g \ h + [\rho_S \ (1 - \epsilon_{FCfb}) + \rho_L \ \epsilon_{FCfb}] \ g \ (L_{RC} - h)$

### Control Valve

(7) $G_S = C \ \phi \ (A_{IP}/A_{FC}) \ (1 - \epsilon_{RCbed}) \ \rho_{av} \ 2 \ (P_{RC0} - P_{FC0}) / \rho_{av}$
where:
C is an empirical efflux coefficient
$\rho_{av} = [\rho_S \ (1 - \epsilon_{RCbed}) + \rho_L \ \epsilon_{RCbed}]$

### Solids Mass Balance

(8) $H = \rho_S \ A_{FC} \ (1 - \epsilon_{FC} (x)) dx + \rho_S \ A_{RC} \ h \ (1 - \epsilon_{RCbed}) + \rho_S \ A_{RC} \ (L_{RC} - h) \ (1 - \epsilon_{RCfb}) + H_{CY} + H_{VA}$

Two persistent recycles appear in the information flow structure of the system, upon suitable information flow reversal (Fig. 2). The system appears to contain 9 variables exchanged between blocks. $P_{FC0}$ is considered twice since it enters into both fast column and valve block. It has been conceived as a recycle variable whose value (together those of $V_{L1}$, $\epsilon_{FC1}$, and $P_{FC1}$ entering into the hydrocyclone block) must be compatible with the condition $P_{FC1} = 1$ bar. The overall system degrees of freedom are the difference between the sum of the local degrees of freedom for each component block and the number of variables connecting the blocks to each other. Therefore, there are 3 degrees of freedom for the whole system, namely: 12 local degrees of freedom minus 9 interconnecting variables ($\epsilon_{FC}(x)$, $G_S$, h, $P_{RC0}$, $P_{FC1}$, $\epsilon_{FC1}$, $V_{L1}$, and $P_{FC0}$, the latter to be considered twice). For the given block assemblage, the three design variables are H, $U_L$ and $\phi$. One recycle variable is h, while the other is indirectly given in terms of $P_{FC0}$.

## RESULTS AND DISCUSSION

Results presented in the following refer to a CFB system whose geometrical parameters are specified in Table 2. Solids are spherical, with a particle density of 1600 Kg/m³ and a particle size of 100 μm. Liquid phase is water.

### TABLE 2.

### CFB SYSTEM GEOMETRICAL PARAMETERS

$D_{FC}$ = 0.04 m

$L_{FC}$ = 3.00 m

$D_{RC}$ = 0.10 m

$L_{RC}$ = 2.70 m

$D_{IP}$ = 0.127 m

$D_{CY}$ = 0.09 m

$L_{CY}$ = 0.30 m

### Voidage profiles

Fig. 3 shows axial voidage profiles along the fast column for different superficial liquid velocities $U_L$ and given values of H and $\phi$. Due to the small size of solids, voidage levels off at a short distance from the entrance. $\epsilon_{FC}(x)$ rapidly increases above 0.95 for $U_L > 0.3$ m/s, which indicates that solids concentration in the fast column is low. Level-off values of voidage $\epsilon_{FC}(x)$ are shown in Fig. 4, for given H, as a function of both $U_L$ and $\phi$. At any given $U_L$, it can be seen how voidage increases by decreasing $\phi$.

### Mass flux of recirculating solids

For any given H, the behavior of $G_S$ as a function of $U_L$ and $\phi$ is better understood when considering the influence of $U_L$ on pressure drop in the four blocks of the system.

Figures 5A, 5B and 6A show pressure drop through the fast column ($\Delta P_{FC}$), the hydrocyclone ($\Delta P_{CY}$) and the recirculation column ($\Delta P_{FC}$), respectively, as a function of $U_L$. The minimum in the $\Delta P_{FC}$ vs $U_L$ curve (Fig. 5A) stems from the contrasting effects produced by an increase in $U_L$. Indeed, increasing $U_L$ increases the voidage in the fast column (thus reducing the gravity pressure drop term in eq. 3), while simultaneously increasing pressure drop due to friction on the wall [Teo and Leung, 1984]. As regards Figs. 5B and 6A, both $\Delta P_{CY}$ and $\Delta P_{RC}$ increase with $U_L$. The former, $\Delta P_{CY}$, increases rather rapidly, since it is directly related to the flow through the hydrocyclone. On the contrary, $\Delta P_{RC}$ increases slowly since it is associated with the increasing liquid flow through the recirculation column. The change in $\Delta P_{RC}$ stems from the increase in the amount of solids contained in this part of the system that, in turn, results from the dilution of the suspension in the fast column.

In Fig. 6B, the pressure drop through the regulation valve is reported as a function of $U_L$. It can be seen that a maximum appears at low $U_L$ values. Under steady state operation, the pressure drop between the points 1 and 2, on the recirculation column side (Fig. 1), is equal to that between points 1 and 2, on the fast column side.

Therefore :

$$(9) \quad \Delta P_{RC} = \Delta P_{VA} + \Delta P_{FC} + \Delta P_{CY}$$

The pressure drop through the valve is:

$$(10) \quad \Delta P_{VA} = \Delta P_{RC} - \Delta P_{FC} - \Delta P_{CY}$$

which is consistent with the fact that ordinates in Fig. 6B are equal to the algebraic sum of the corresponding ordinates in Figs. 5A, 5B and 6A, according to Eq. 10.

$G_S$ is reported in Fig. 7, as a function of $U_L$ and $\phi$, with H = 30 kg. For any given $\phi$, $G_S$ varies with $U_L$ according to a relationship that is similar to that involving $\Delta P_{VA}$ and $U_L$. In particular, for any $\phi$, a maximum of $G_S$ is found at low values of $U_L$. The similarity between curves reported in Fig. 6B and in Fig. 7 is simply explained as a consequence of the dependence of $G_S$ on $\Delta P_{VA}$.

It should be noted that, for any given $\phi$, there is a limiting value of $U_L$ at which $G_S$ becomes zero. This corresponds to the condition that the static head related to the presence of any given amount of solids H in the recirculation column balances the pressure drop across the hydrocyclone plus that due to the frictional resistance to liquid flow in the fast column. Figure 8 shows that decreasing H to 20 kg reduces both the mass flux of solids and the limiting liquid superficial velocity (for which $G_S$ = 0).

Comparison with air operated CFB systems. The comparison between model results for CFB systems operated with water and with air is rather instructive. It can be performed by taking into account the results shown in Figs. 3 to 8 and those obtained when air is the fluidizing agent, both in the fast and in the recirculation column [Arena et al., 1987].

The main effect of the change is that, being $(\rho_S - \rho_L)$ smaller than $(\rho_S - \rho_G)$, the difference between static head upstream and downstream the valve is smaller for water than for air. As a consequence, the values of $\Delta P_{VA}$ obtained when the recirculation and the fast column are both filled up with solids-water suspensions are smaller than those produced with solids-air suspensions, which involves higher voidages $\epsilon_{FC}(x)$ and smaller solids mass flux $G_S$.

Such disavantages are limited by increasing the cross section of the recirculated solids inlet at the bottom of the fast column. Provided the intersection between inclined pipe and fast column is correctly designed to avoid any local restrictions to flow, increasing $A_{IP}$ results in a corresponding decrease in $\epsilon_{FC}(x)$ and in an increase in solids mass flux through the CFB system.

## CONCLUSIONS

The work shows the hydrodynamic feasibility of a liquid operated CFB system to be used in anaerobic fermentations.

Model results indicate that the fast column tends to work at voidages that might be somewhat higher than those required by fermentation processes. As far as the entire loop is concerned, recirculating solids mass fluxes are achieved that could be relatively low as compared to continous fermentation requirements.

A correct design of the CFB system as regards solids inlet at the bottom of the fast column can overcome these drawbacks.

## ACKNOWLEDGEMENT

Authors are indebted to Dr. U. Arena for critical reading of the paper.

## NOTATION

A, cross sectional area, m²
C, empirical efflux coefficient, -
D, diameter, m
$F_S$, drag force per unit volume of particle, Pa/m
$F_W$, pressure gradient caused by wall friction in fast column, Pa/m
$F_{WL}$, contribution to $F_W$ due to liquid, Pa/m
$F_{WS}$, contribution to $F_W$ due to solids, Pa/m
g, acceleration due to gravity, m/s²
$G_S$, solids mass flux in fast column, kg/(sm²)
h, expanded bed height in recirculation column, m
H, solids hold-up, kg
L, height, m
P, pressure, Pa
$U_L$, superficial liquid velocity, m/s
V, actual velocity, m/s
x, height coordinate, m

## Greek Symbols

$\Delta P_{FC}$, pressure drop through fast column, Pa
$\Delta P_{RC}$, pressure drop through recirculation column, Pa
$\Delta P_{CY}$, pressure drop through hydrocyclone, Pa
$\Delta P_{CY}$, pressure drop through solids control valve, Pa
$\epsilon$, voidage, -
$\rho$, density, kg/m³
$\rho_{AV}$, average density of liquid-solids suspension in fast column, kg/m³
$\phi$, solids control valve opening, -
b, e, n, $\gamma$, O, Q, $f(Re_{in})$, $C_{in}$, parameters of hydrocyclone design equation [Rietema, 1961]

## Subscripts

bed, fluidized bed in recirculation column
CY, hydrocyclone
fb, free-board of recirculation column
FC, fast column
G, gas
IP, inclined pipe
l, top of the column
L, liquid
o, bottom of the column
RC, recirculation column
S, solids
VA, valve

# REFERENCES

- Arena, U., L. Massimilla, and D. Pirozzi (1987). "Analysis of a Circulating Fluidized Bed System", in <u>FBC comes of age</u>, ed. J.P. Mustonen, 1987 Int. Conf. of FBC, The American Soc. of Mechanical Eng., pp. 762-769.

- Basu, P. (editor), (1986). <u>Circulating Fluidized Bed Technology</u>, 1985 First International Conference on Circulating Fluidized Beds, Pergamon Press, New York.

- Basu, P. (editor), and J.F. Large (editor), (1988). <u>Circulating Fluidized Bed Technology II</u>, 1988 Second International Conference on Circulating Fluidized Beds, Pergamon Press, New York.

- Massimilla, L., G. Volpicelli, and F, Zenz (1963). "Flow of fluid-particle suspensions from liquid-fluidized beds", <u>I&EC Fundamentals</u>, No. 3, pp. 194-199.

- Rietema, K. (1961). "Performance and design of hydrocyclones (I-IV)", <u>Chem. Eng. Science</u>, No. 15, pp. 298-325.

- Rudd, F.D., and C.C. Watson (1968). <u>Strategy of Process Engineering</u>, Wiley, New York, pp. 34-79.

- Soo, S.L. (1967). <u>Fluid Dynamics of Multiphase Systems</u>, Blaisdel Publish. Co., Waltham, MA.

- Teo, C.S., and L.S. Leung (1984). "Vertical Flow of Particulate Solids in Standpipes and Risers", in <u>Hydrodynamic of Gas-Solids Fluidization</u>, Eds. N.P. Cheremisinoff and P.N. Cheremisinoff, Gulf Publish. Company, Houston, Texas, pp. 471-542.

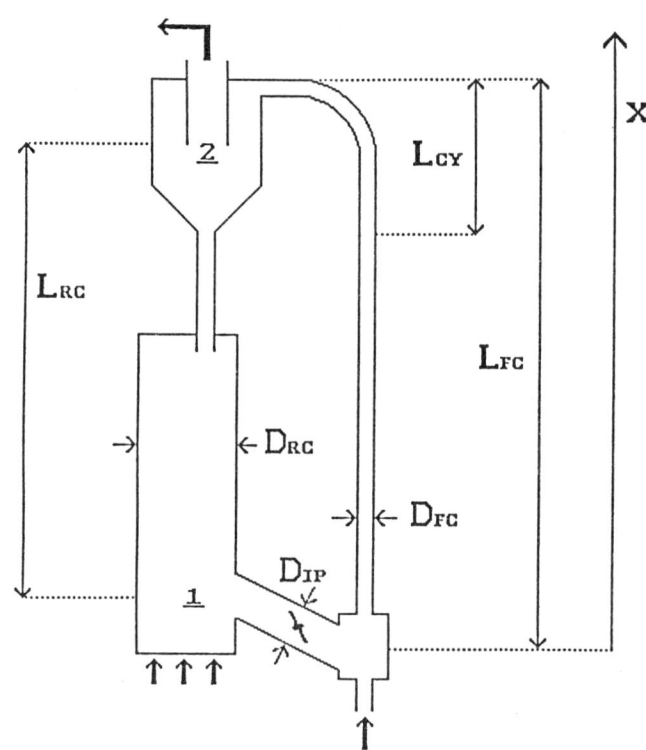

Figure 1. Schematic representation of the CFB system. The meaning of the geometrical parameters of the system is in the nomenclature. 1 and 2 are reference points for pressure drop determination (eqns. 9 and 10). Characteristic lengths of the hydrocyclone are as in Rietema (1961).

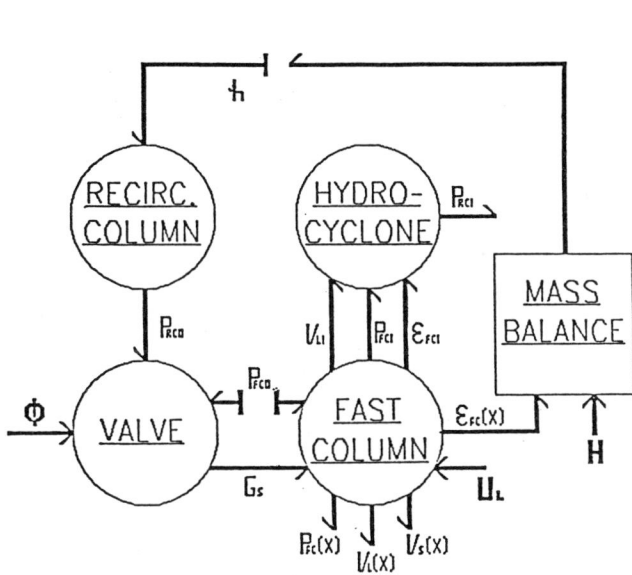

Figure 2. CFB system information flow structure.

→ design variable
⇒ state variable
⊢ recycle variable

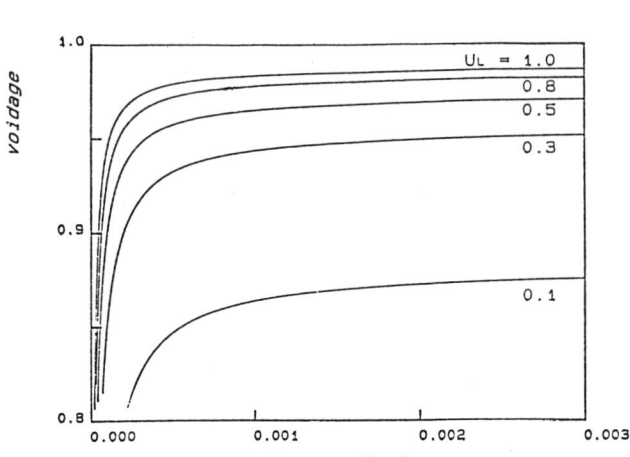

Figure 3. Axial voidage profiles along the fast column. H = 30 kg, $\phi$ = 0.5.

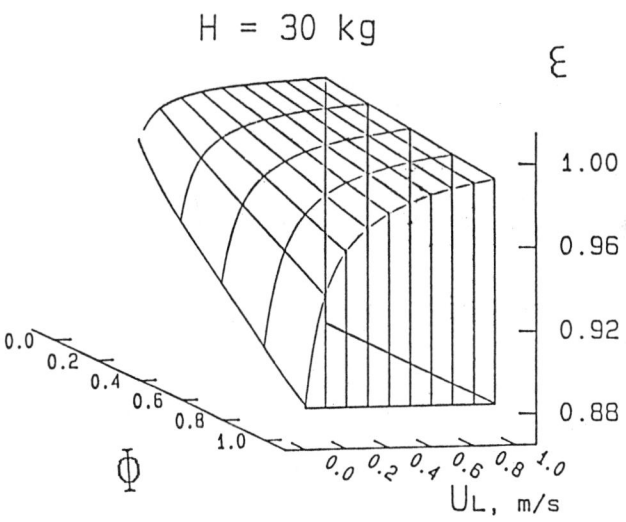

Figure 4. Spatial representation of level-off values of $\epsilon_{FC}(x)$ as a function of $U_L$ and $\phi$. H = 30 kg.

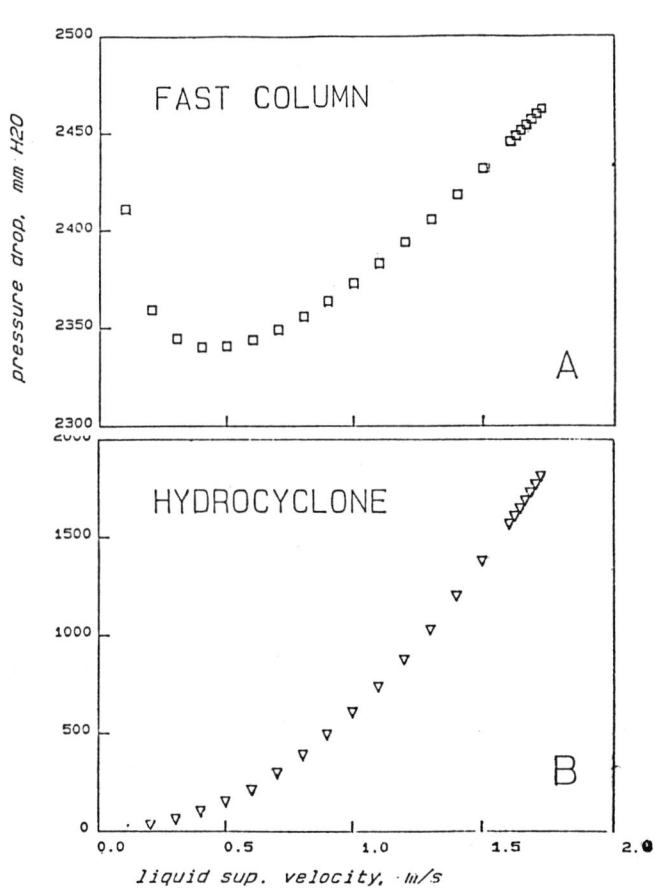

Figure 5. Pressue drop through the blocks of the system as a function of $U_L$: Fast column (A) and Hydrocyclone (B). H = 30 kg, $\phi$ = 0.1.

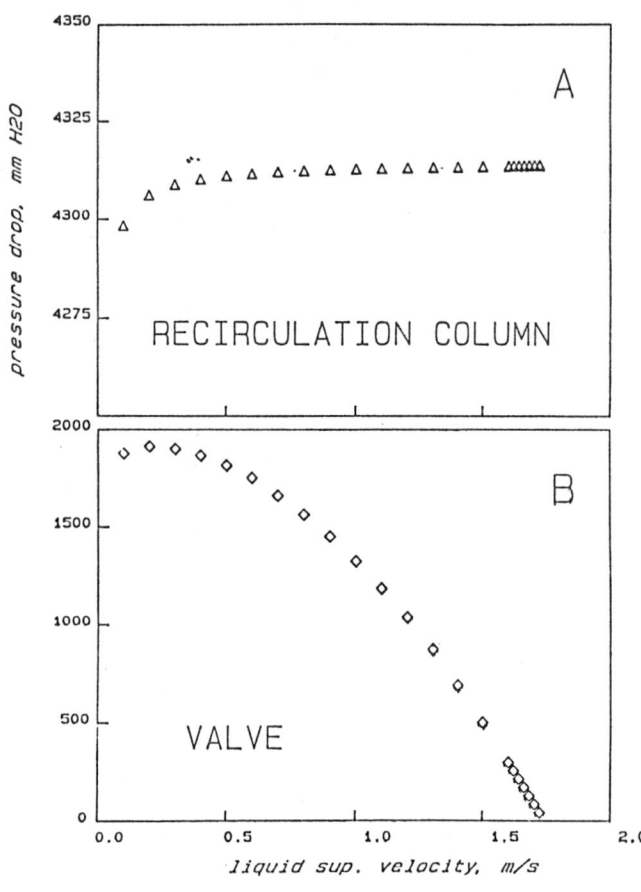

Figure 6. Pressue drop through the blocks of the system as a function of $U_L$: Recirculation column (A) and Valve (B). H = 30 kg, $\phi$ = 0.1.

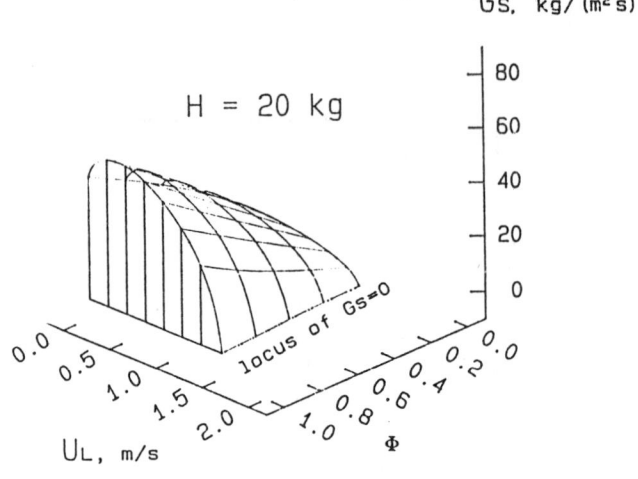

Figure 8. Spatial representation of $G_s$ as a function of $U_L$ and $\phi$. H = 20 kg.

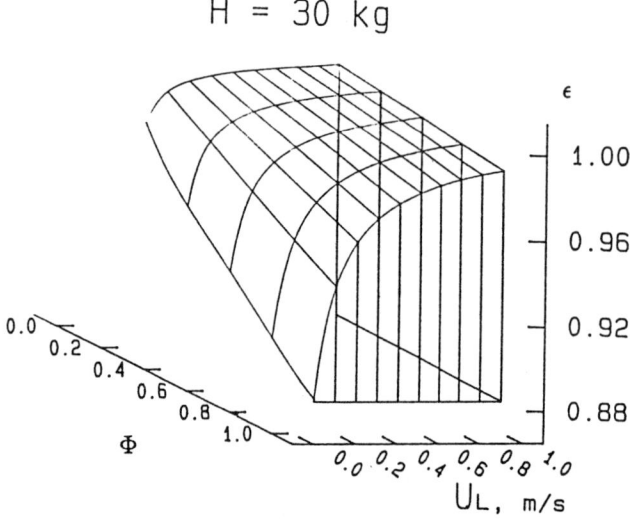

Figure 7. Spatial representation of $G_s$ as a function of $U_L$ and $\phi$. H = 30 kg.

# CHALLENGES IN FLUIDIZED BED TECHNOLOGY

Derek Geldart ■ Postgraduate School of Powder Technology, Department of Chemical Engineering, University of Bradford, West Yorkshire, UK

Although the technologies are almost 50 years old, fluidization and the flow of aerated powders still present many scientific, engineering and conceptual problems. Because many of the basic parameters in powder systems are not easily defined, few of the operational difficulties encountered in industrial units can be solved using the 'high-tech', sophisticated computer models which are currently fashionable in chemical engineering. Experimental work with powders is still essential but is often perceived as 'low-tech', dirty and unglamorous by young researchers. Nevertheless, the planning of experiments, interpretation of results and their use in scaling up processes offer a range of difficult intellectual challenges to engineers in the academic and industrial worlds.

Some of the important unsolved or poorly understood mechanisms involved in fluidized bed catalytic reactors and other aerated fine powder systems are presented and discussed, in the hope that they will attract the attention of enthusiastic experimenters.

The story of the successful development of Fluid Bed Catalytic Cracking has been told on a number of occasions recently, both by some of those who were directly involved in the early 1940's (e.g. Jahnig) and also by others with a more general interest in the technology (e.g. Squires, Spitz) and it would not be appropriate to repeat the tale here, fascinating though it is. Other equally successful fluid bed processes have received much less exposure probably because fewer were built and fewer contractors were involved - acrylonitrile manufacture is one and oxychlorination of ethylene to give vinyl chloride is another. Worldwide there are at least 450 large fluid bed units operating these three processes, of which 350 are crackers. New organic synthesis processes using catalysts are reported each year, and although few reach successful large scale production, all have in common the fact that porous catalysts are used which have mean sizes in the range 45-75 $\mu$m and particle densities 1000-1800 kg/m$^3$ - what I have called Group A powders. The mean size, density, and the level of 'fines', usually characterised by the weight percentage less than 45 $\mu$m, are all known to influence both the flow properties of the catalyst and the chemical conversion. The sphericity of the particles is also thought to be important. However, in spite of the fact that the three production processes have been operated for 30-40 years, we still do not understand with any certainty the mechanisms by which the parameters listed above influence the behaviour of these very important pieces of chemical plant. Moreover, the ways in which the various mechanical components in the system behave and interact with each other and with the powder are poorly understood.

This deficient understanding, though frustrating (especially to the academic!), could in itself be tolerated by management and industrial engineers since on the whole the existing processes work well enough - production managers are strong believers in "If it aint't broke, don't fix it!". Modifications to large units based on hunches or incomplete understanding can (and have) proved to be disastrously expensive if they fail.

However, there are several good reasons why research work should be done in order to improve our understanding of these processes:

(i) small improvements in conversion and/or yield - as little as 0.5% - could increase profits by millions of dollars worldwide.

(ii) smoother running of existing units could be better achieved when the composition of the catalyst or of the feed is changed.

(iii) start-up and running of new units which may be smaller, larger or of slightly different design would be less likely to cause problems.

(iv) the increasing manufacture of industrial ceramic materials in aerated, pneumatic and fluid bed processes means that finer and more cohesive powders (Group C) need to be handled and their behaviour is notoriously difficult to predict. This makes it difficult to design reliable equipment.

Thus there is a real need to understand properly the behaviour of powders in Groups A and C and the plant in which they are fluidized since this can open the way to new processes and products in the 21st century. These problems offer considerable intellectual and experimental challenges but, because they are not easy to formulate, they cannot yet be expressed mathematically nor solved in powerful and ingenious computer simulation programs. Added to this is the fact that experimental work with powders is perceived to be dirty, 'low-tech' and unglamorous; and, powder technology figures hardly at all in most undergraduate or graduate courses in the USA. It is not surprising therefore that young academics raised on liquids, gases and computers find the field unappealing.

I shall now detail some of the fascinating industrial problems which require attention, in the hope that this will attract the attention of enthusiastic experimenters.

SOME PROBLEM AREAS IN LARGE CATALYTIC REACTORS.

Figure 1 shows a typical unit in which an organic synthesis using a Group A catalyst might take place. Reactor diameters are in the range 1-10 m, gases A and C may be one and the same or different, there may or may not be a tube bank, and catalyst feed and removal may be considerable and continuous, or small and infrequent.

Region 1 - Freeboard. (Fig. 2)

Significant conversion may occur here and this is influenced by the axial and radial solid concentrations. How these are influenced by gas velocity, particle properties, equipment size, internals such as heat transfer surfaces and cyclone barrels and diplegs, and particle properties is not properly understood. The entrainment rate determines not only the density profiles in the freeboard, but also the rate of recycle of fine solids back to the bottom of the reactor and cannot yet be predicted reliably.

Region 2 - The dense region or 'bed'. (Fig. 3)

Much of the conversion takes place here but what is the mode of fluidization and how does the gas/solid contact occur? Are we justified in thinking in terms of bubbles and equilibrium sizes or is the flow regime chaotically 'turbulent'? What are the vertical and radial density profiles and, as in the freeboard, how are they affected by column diameter, internal surfaces and particle properties? What are the solids flow patterns in large reactors with and without internals and how are they affected by solids recycled down diplegs, fresh catalyst pneumatically injected or equilibrium catalyst removed near the distributor? How do these factors influence temperature profiles, and how does the level at which solids are reinjected influence conversion?

Region 3 - Grid Region. (Fig. 4)

For fast reactions most of the conversion takes place near the grids. How can even distribution be ensured over a column 5-10 m diameter? If the reaction is exothermic how can good temperature control be achieved in a small vertical region of the bed whilst at the same time avoiding erosion of heat transfer tubes? What is the rate of mass transfer to and from a jet in a fluid bed? How can good gas/gas contact be achieved between gases A, B and C?

Region 4 - Flow Down Standpipes and Diplegs. (Fig. 5)

The solids flowing down cyclone diplegs have differing properties; those in the first stage are similar to the bed material and pass downwards at high velocity, whilst those captured in the third or fourth stage are cohesive and discharge discontinuously. What is the flow regime in the first stage dipleg and is it uniform vertically and radially. Is there a recycle of product gas down this dipleg caused by the massive downwards flow of solids and how much?

How does this gas flow affect gas and solids flow patterns near the grid, and the conversion and yield? How do trickle valves on the secondary and tertiary dipleg work? What happens when they open? Do gas and solids rush in? Is there a continuous leakage of gas up the dipleg before the pressure balance is satisfied, and if so, does it affect conversion and cyclone efficiency? What is known about particle attrition in the cyclone and how can it be reduced?

Are electrostatics important in catalytic reactors and if so, how?

If, after 40 years, we don't know the answers to these questions involving 'ideal' manufactured powders how well can we deal with fluidized systems involving the manufacture and use of much finer less well-defined materials? Some of the answers may well be found by experimental work in cold models, but the problems of scale-up from cold to hot, and small to large, are always there. In the long run, a proper understanding of the fundamental mechanisms is required if basic design and scale-up are to be done without expensive mistakes.

In the next section I focus more closely on some of these questions and in particular on fundamentals.

POWDER BEHAVIOUR

- based on laboratory experiments and industrial experience

Let us remind ourselves of the aerated properties of powders in Groups A and C which constitute the catalysts used in fluid bed catalytic processes (see Figure 6). Beds of powder in Group A expand considerably at velocities between minimum fluidization $U_{mf}$, and minimum bubbling, $U_{mb}$ (Figure 7). At velocities above the minimum bubbling velocity, bubbles grow rapidly with velocity and with distance above a porous distributor (Figure 8A). Conversely, large bubbles formed from pipe grids having large widely-spaced holes may break up (Figure 8B). Eventually an equilibrium bubble size or 'maximum stable' size is observed in cold models fluidized at velocities up to about 20 cm/s. Many mathematical models of chemical conversion are based on the idea of bubbles, and their protagonists maintain that it is the size of the bubbles and the contact between the gas in the bubbles and the catalyst which determines chemical conversion. However, we do not know for certain whether bubbles still exist in large units operated at 60-80 cm/s at high temperatures.

When the gas supply is suddenly cut off, the bed collapses (or de-aerates) slowly at a rate comparable to the minimum fluidization velocity. This makes Group A powders ideal to circulate around fluidized and pneumatic conveying loops; however, the ease with which they become, and remain, aerated makes them liable to flood on discharge from storage hoppers.

Powders in Group C are cohesive and difficult to fluidize. Because the mass of each particle is small, the interparticle forces are comparable with, or greater than, the gravitational forces. These interparticle forces (electrostatic, van der Waals, vapour and liquid adsorbed layers) are dependent on the size, shape, roughness, hardness, and chemical structure of the particles as well as on the gas properties.

How are all these aerated flow properties influenced by the particle and gas properties, and why? Although we shall be concerned mostly with the 'why', we must first review what is known about the 'how'.

Particle density:

The average FCC unit loses approximately 5 tonnes catalyst per day at a cost of up to $2000 per tonne, 7 days per week, 52 weeks per year. These losses arise largely from attrition of the catalyst. The hardness of the catalyst has a direct bearing on attrition and the response of the catalyst manufacturers has been to make catalysts harder.

In doing so, catalysts have been made denser and this has affected adversely the ability to circulate the solids. Increasing particle density is known to reduce the degree of non-bubbling expansion, to increase the de-aeration rate (Fig. 9), and also to increase bubble size. These together may be the cause of different types of catalyst circulation problems (see Figure 10) in that:

(a) more rapid de-aeration in a tall standpipe may result in defluidizaton.

(b) bubbles pulled down from the hopper at the entrance to the will be larger and occupy a larger proportion of the standpipe cross-section, thus choking off the downwards flow of solids.

Particle size:

Mean particle size has a profound effect on the fluidization behaviour of powders within Groups A and C. An increase in particle size reduces both non-bubbling and bubbling bed expansion, and increases de-aeration rate (Fig. 9); it also increases equilibrium bubble size. There is also some tenuous evidence that bigger particles are more subject to attrition. Certainly, catalyst circulation becomes more difficult as mean particle size increases. Conversely, if mean particle size decreases to below 20-25 $\mu$m, that is the powder behaves as a Group C powder, the degree of cohesivity increases rapidly with further decrease in particle size, and flowability, bed expansion, solids mixing and heat transfer decrease (Figs. 11 and 12). Entrainment rates increase as mean particle size decreases with Group A, but within Group C entrainment decreases with decrease in particle size (Fig. 13).

Particle size distribution:

There is little evidence that the distribution of particle size about the mean influences fluidization behaviour; however, added fines (that is, material less than 45$\mu$m) certainly do influence beneficially fluidization behaviour over and above the effect they have on the mean size of the powder. It is also known that added inert fines can improve the chemical conversion in certain catalytic reactors, clearly indicating a hydrodynamic effect. Bed expansion is increased (Fig. 14), de-aeration times increased and, probably, bubble sizes reduced. The entrainment rate of coarse particles is significantly increased by the presence of finer particles (Fig. 15).

Particle shape:

Largely because of the processes by which commercial catalysts are manufactured, the particles are spherical or well rounded. Few systematic studies have been done in which only particle shape was changed; nevertheless, it is firmly believed, with some anecdotal evidence, that spherical smooth particles give 'better' fluidization than rough, angular particles.

Gas properties:

An increase in operating pressure is well known to give smoother fluidization, largely through smaller bubbles. The effect of different gases on fluidization has generally been interpreted in terms of gas viscosity - higher viscosity gives higher bed expansion and larger de-aeration times - but there are good reasons for believing that gas adsorption on the particle may also play an important role.

FUNDAMENTALS

- based on some experiments and much speculation

Maximum stable bubble size and dense phase voidage: the concept of a maximum stable bubble size goes back to the late 1950's and implies not that bubbles reach a certain size and then remain unchanged but rather that a state of dynamic equilibrium is reached in which splitting and coalescence occur continually. There is little doubt that in Group A powders fluidized at low velocities (less than 0.2 m/s) bubbles do reach an equilibrium size but views on the mechanisms differ and involve differing concepts of the structure of the dense phase. In Group B, powders begin to bubble at or slightly above $U_{mf}$ with virtually no dense phase expansion - bubbles grow indefinitely as bed depth and/or gas velocity are increased.

In Group A, powders fluidize at $U_{mf}$ and exhibit bubble-free expansion until $U_{mb}$ is reached - the smaller the size and density of the particles, the bigger the expansion, and, at gas velocities above $U_{mb}$, the smaller the equilibrium bubble size. Is the latter a consequence of the former? One tentative and qualitative theory for the maximum stable bubble hinges on the mode of splitting of the bubbles and on the degree of expansion of the dense phase. Bubbles in viscous liquids break up by the formation of an indentation which forms on the upper surface of the bubble and grows as it is swept around the periphery. If the indentation grows sufficiently to reach the floor of the bubble before being swept away, the bubble will split. When surface tension effects are negligible, as in fluidized beds, the growth rate of a disturbance increases with decreasing kinematic viscosity of the surrounding medium. This mode of bubble splitting has been observed in two-dimensional fluidized beds, but is this typical of what happens in a three-dimensional bed?

Some workers believe that in three-dimensional beds, bubbles split from the bottom. What experiments can be done to verify this? Is any of this relevant to what happens in a reactor fluidized at 60 cm/s when the system may no longer resemble a matrix of bubbles within a continuous phase of aerated powders, but rather clusters or clumps of particles within a continuous gas phase? Is this what is meant by the term 'turbulent fluidization'?

It is now well established that when small particles are circulated through a fluidized bed of large (non-entrainable) particles, the transport disengagement height of the latter is increased substantially. Indeed, if the entrainment flux of the small particles is sufficiently high, the large particles may be carried out of the column altogether, that is, they become entrainable. How can this be modelled? I believe that the fine particles collide with the larger ones flung into the freeboard by the action of bubbles, give up some or all of their momentum and are then re-entrained by the gas. The large particles are impelled further up the column by this action, the height to which they are carried depending on their size and the number of interparticle collisions. A mathematical model based on this concept has yet to be formulated; in the meantime, a correlation based on the idea of an "effective density" of the gas/fine particle mixture works well in some cases but can lead to absurd overprediction in others.

Is the basic concept correct and can it be modelled? There is certainly a considerable challenge here.

## The rôle of interparticle forces

The idea of a viscosity of a fluidized bed analogous to that measured in a liquid has been in the literature since the late 1940s.

Certainly dense phase 'viscosity' decreases with bed expansion, but why does the dense phase expand? One school of thought (Rietema and co-workers) attributes the expansion with decreasing particle size to the presence of small interparticle forces; another group (Foscolo, Gibilaro, et al.) believes that the phenomenon can be explained in purely hydrodynamic terms. If interparticle forces are the cause, which forces are important and how are they affected by pressure and temperature? If purely hydrodynamic forces are involved, why should bubbles in a bed of spherical Group A powder be smaller than those in a bed of angular particles of similar size, as found in some recent work?

The structure of the dense phase will be very different depending on whether interparticle or purely hydrodynamic forces are involved. In the former case we can picture the particles as being in contact at all times as though attached to each other with chewing gum; tiny gaps open up at random and allow gas to percolate through (Fig. 16A). When the chewing gum becomes too extended and weak under the influence of fluid drag, a particle separates from one neighbour and moves closer to another. The hydrodynamic model pictures particles entirely surrounded by fluid and separated more or less evenly from each other, except when a continuity wave moves through the system (Fig. 16B). Such waves certainly occur in liquid fluidized systems in which interparticle forces are largely absent. Particles are too small to be seen in the centre of a bubbling bed but what experiments can be devised to decide which model is correct?

Some recent Japanese work on a co-grinding process called 'mechanofusion' demonstrates clearly the importance of the nature of the particle surface on powder flow. Particles are ground up with small amounts of another much finer material, which is thereby forced into the surface layers of the parent particles. The result can be that two very cohesive powders can form a new material which will flow freely and can be fluidized. Clearly hydrodynamic forces play no rôle here - the magnitude of the forces between particles dominates completely.

It is well known that tiny amounts of ultra-fine particles such as Cab-o-sil, Alusil and others can promote the fluidization of otherwise cohesive powders yet the reasons are obscure. In cases in which the cohesivity is caused by adsorbed moisture, it seems likely that this is mopped up by the immense surface area of the ultra-fine particles. In other cases, where water is absent, flow improvement certainly cannot be explained so simply. There is clearly scope and need for some fundamental work in this area.

## CONCLUSIONS

I have confined my discussions largely to the fluidization of fine particles but there are many other intriguing and important questions to do with coarse particles which remain to be answered. The entire question of scaling based on non-dimensional groups has become a hot topic — especially when applied to fluid bed combustion! I hope that I have been able to show young researchers especially that in spite of the voluminous literature there is still much to be discovered in fluid bed technology. Don't be discouraged by the dirt and the dust — get in there and experiment! The field is still full of surprises so — good fluidizing!

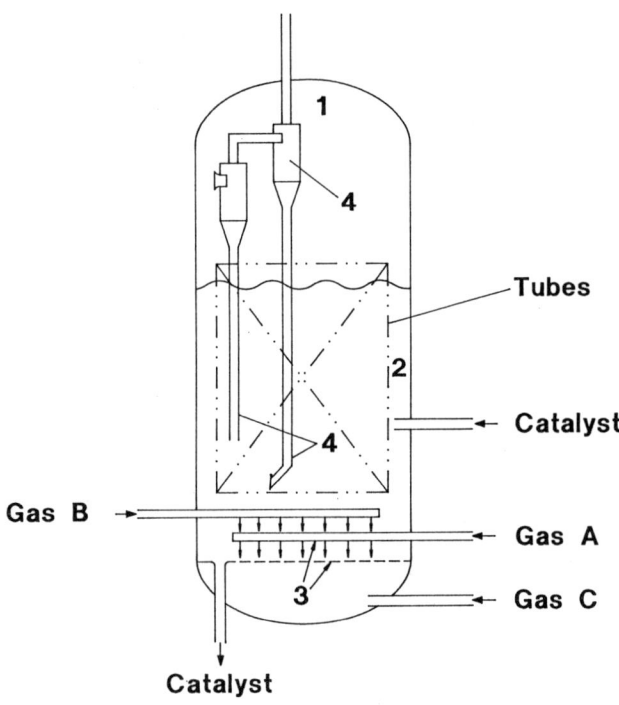

Figure 1. Typical fluid bed catalytic reactor.

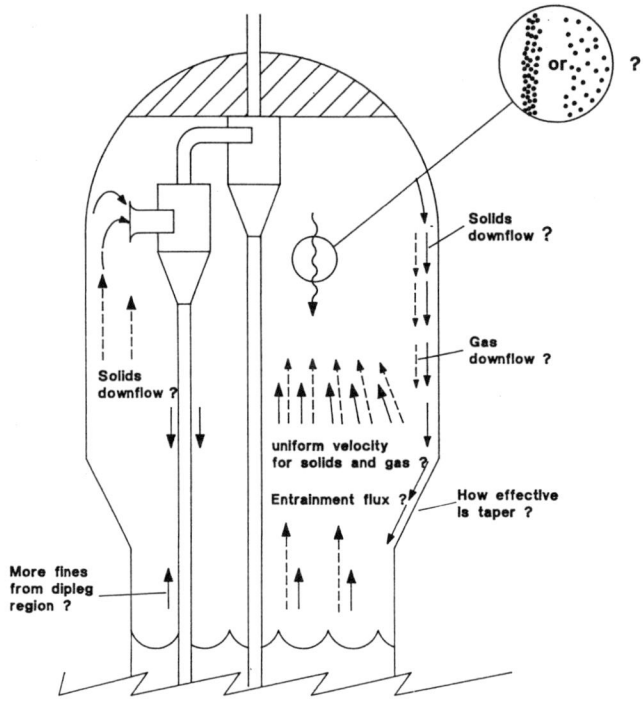

Figure 2. Freeboard region.

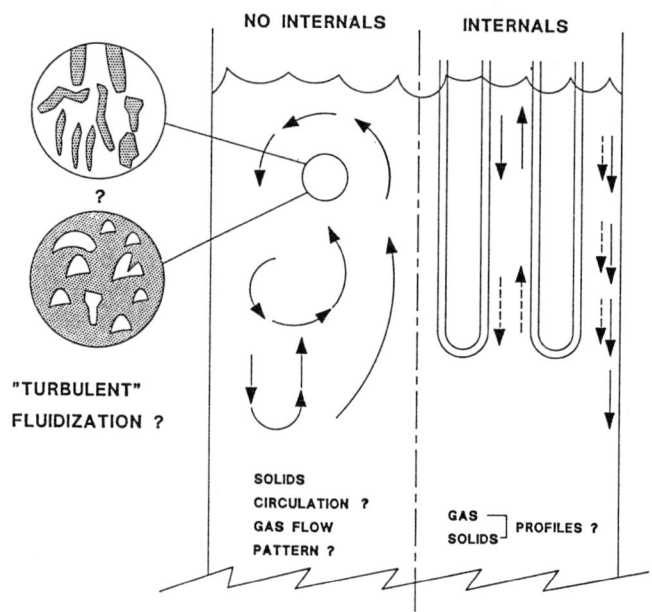

Figure 3. "Bed" region.

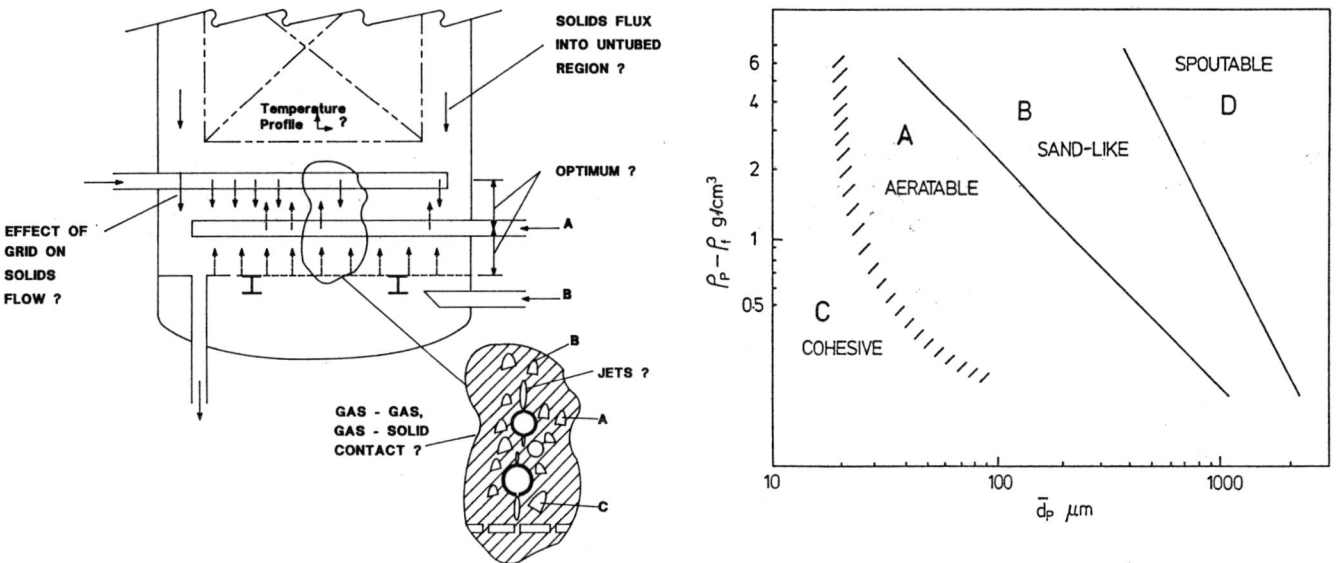

Figure 4. Grid region.

Figure 6. Powder classification diagram.

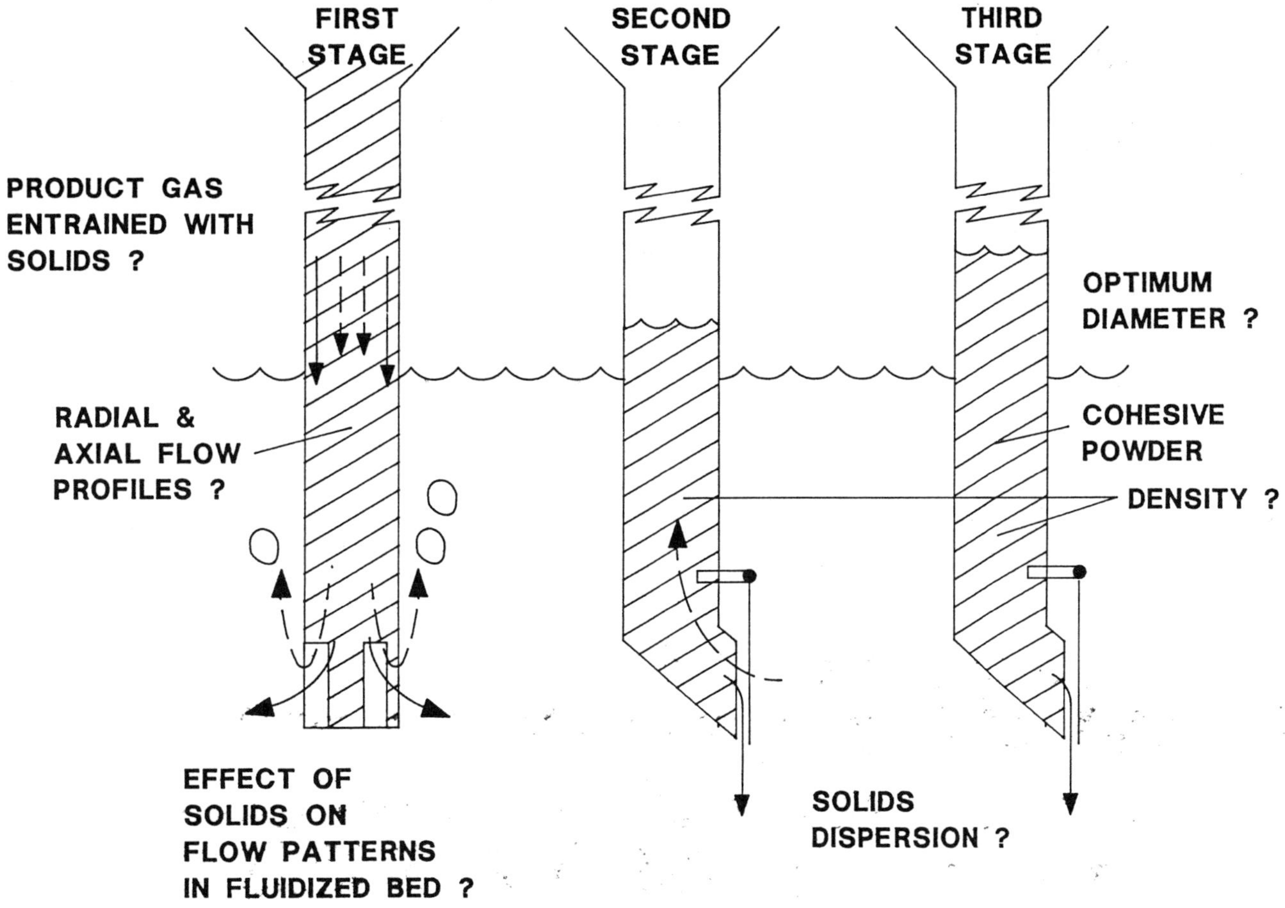

Figure 5. Standpipes and diplegs.

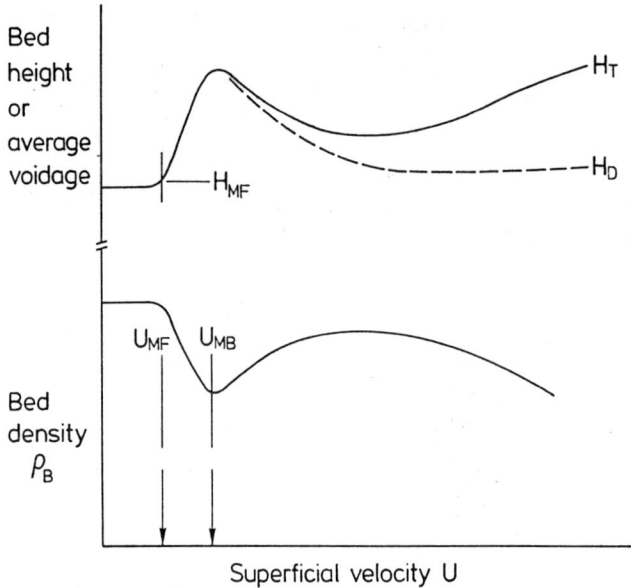

Figure 7. Bed expansion of a group A powder.

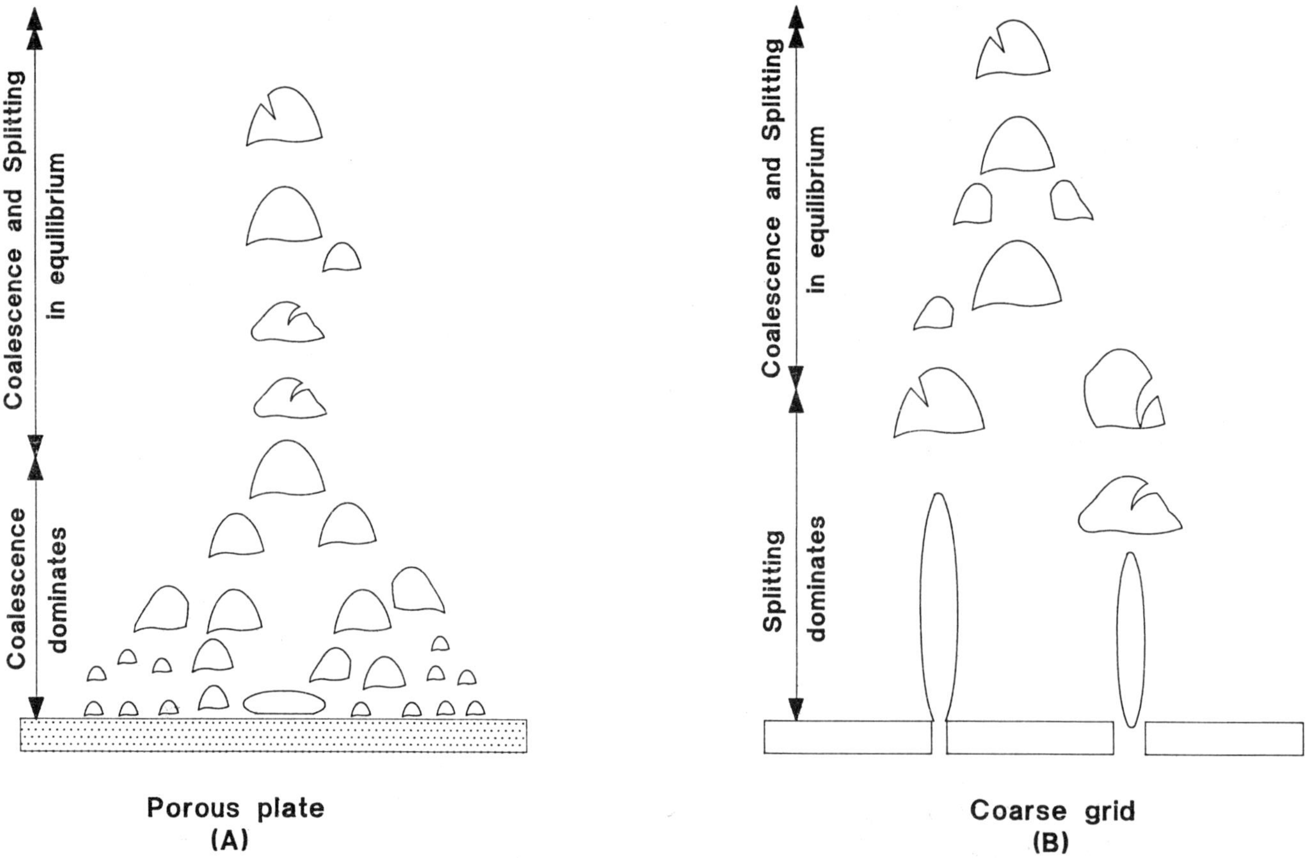

Figure 8. Bubbles reach equilibrium size whatever the grid design.

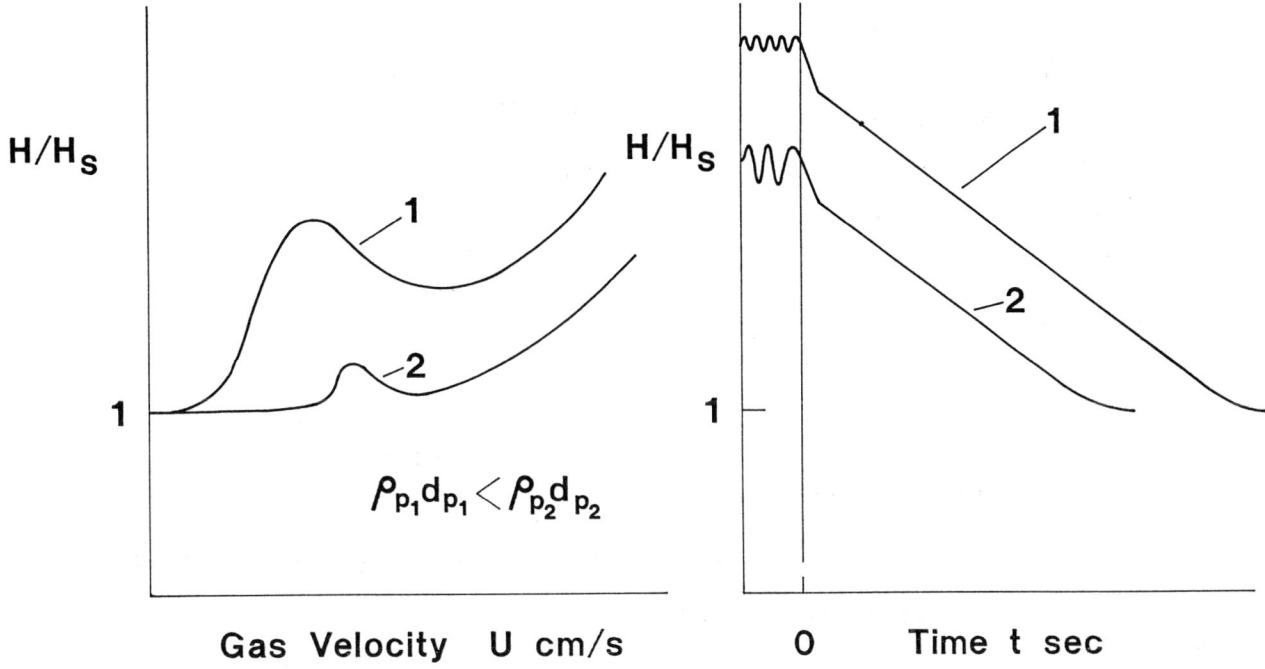

Figure 9. Effect of particle size and density on bed expansion and de-aeration.

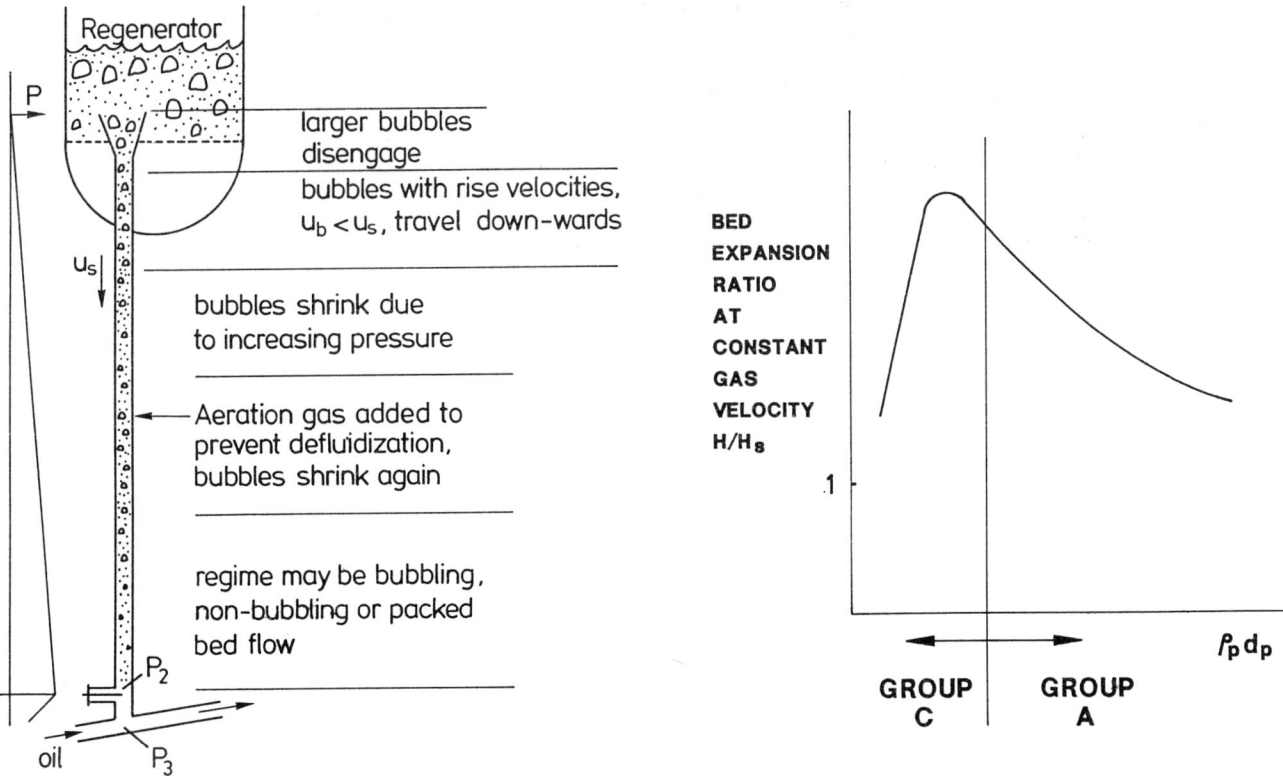

Figure 10. Zones in a long standpipe.

Figure 11. Effect of particle size and density on bed expansion ratio.

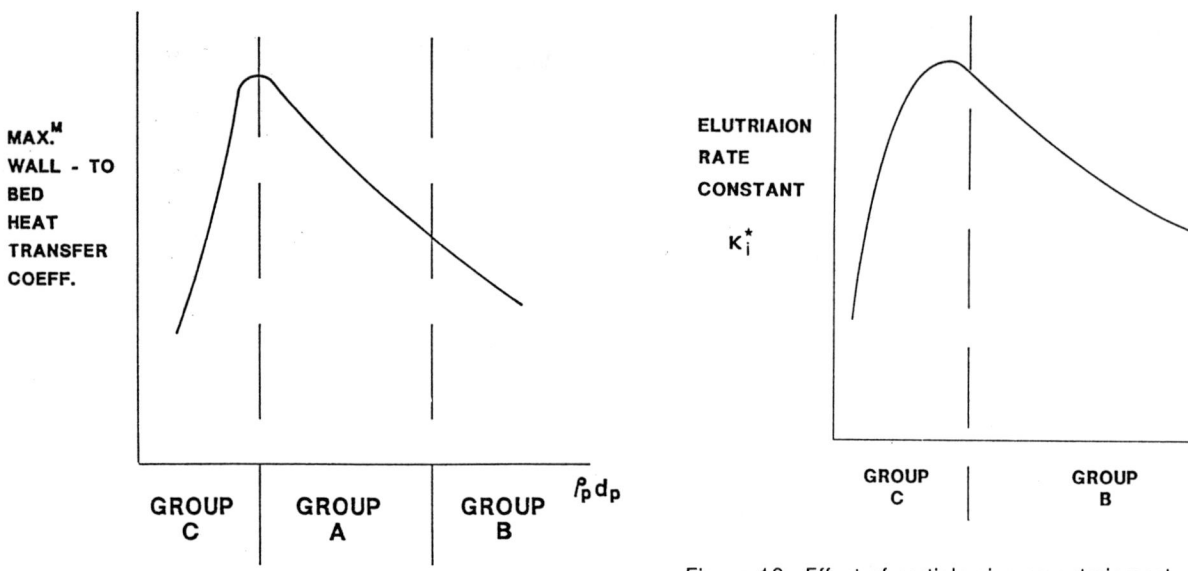

Figure 12. Effect of particle size and density on wall-to-bed heat transfer coefficient.

Figure 13. Effect of particle size on entrainment.

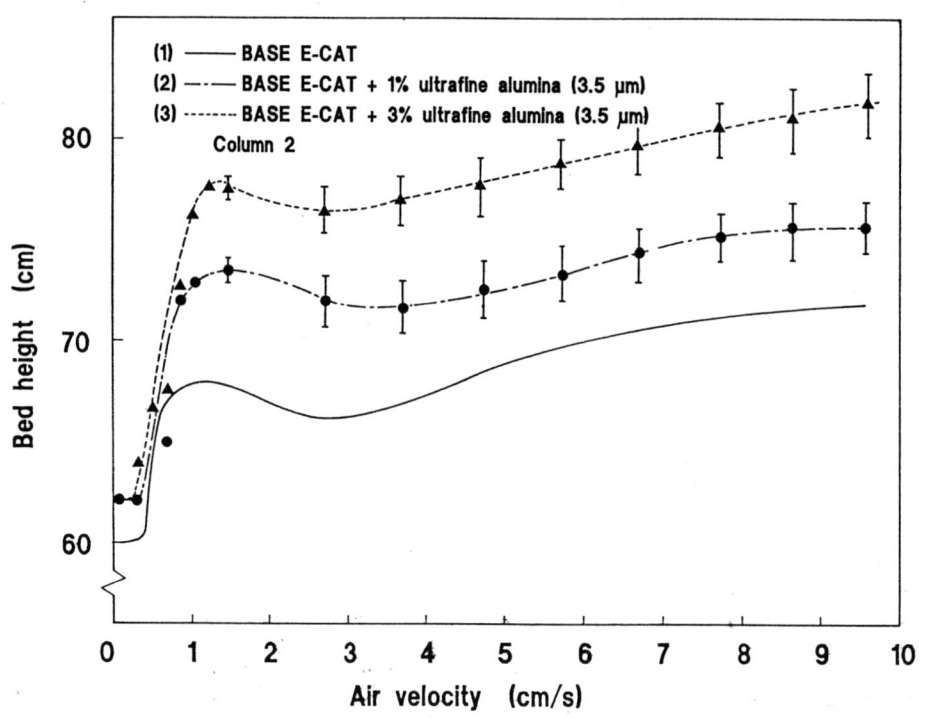

Figure 14. Addition of small amounts of very fine particles increases bed expansion.

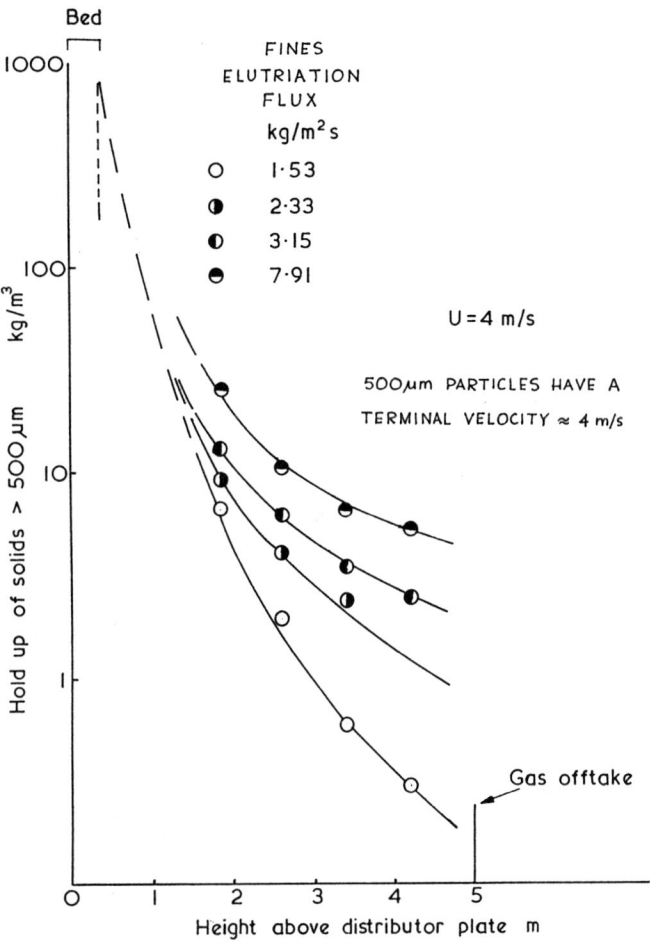

Figure 15. Influence of fines on elutriation of coarse particles.

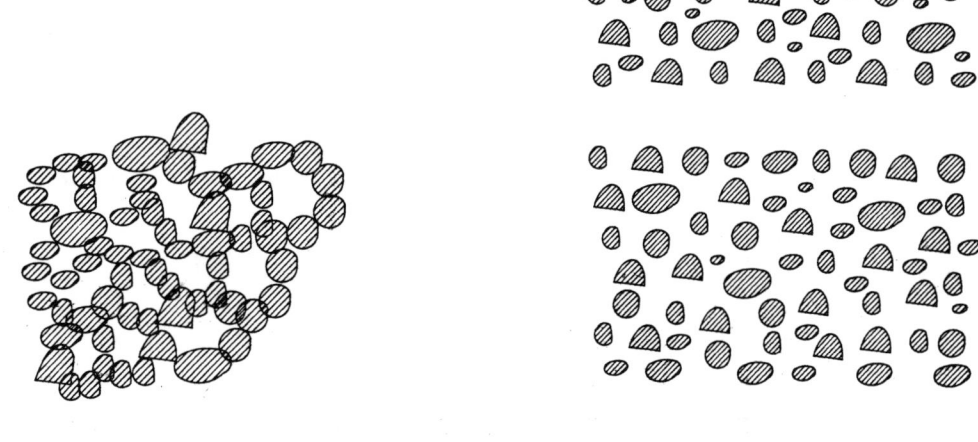

A. Interparticle forces present

B. Only hydrodynamic forces present

Figure 16. Two concepts of the structure of the dense phase in a group A powder.

# INDEX

## A
acidic gases, removal of .......................... 94
aerosol processes, material synthesis by ................ 57

## B
bed, transients in ............................ 49
bubbles, computer simulation of .................... 22
bubbling, pressure on transition from ............... 37

## C
catalytic cracking, ultra-rapid ..................... 69
circulating fluidized bed fermentor, development of .... 101
cohesive powder structures, tensile strength of ........ 44

## D
decomposition, silane reaction ..................... 77
dry-scrubbing process, laboratory testing of .......... 94

## F
fines generation, effect of system parameters on ....... 83
flow conditioners, effect of ...................... 44
flue gas, simulated incinerator .................... 94
fluidization, pressure from bubbling to turbulent ....... 37
fluidized bed,
    bubbles in .................................. 22
    circulating liquid ............................ 32
    circulating ................................. 101
    expansion of a three-phase ..................... 49
    laboratory testing of .......................... 94
    particle mixing in ............................ 32
    reactor, elutriation in ........................ 77
    technology, challenges in ..................... 111
fluidized limestone/coal-char mixtures ............... 83

## G
gas-solid trickle flow, hydrodynamics of ............. 11

## H
heavy oils, ultrapyrolysis of ...................... 69
hydrodynamic model for fluidized bed
    fermentor system ........................... 101

## L
limestone/coal-char mixtures ...................... 83

## M
material synthesis, an overview of .................. 57
mixtures, fluidized limestone/chal-char ............... 83

## O
operating temperature, effect of .................... 37

## P
pneumatic conveying systems, at various inclinations ..... 1
pressure, effect of operating temperature ............. 37

## R
reactors, ultra-rapid fluidized ..................... 69
regularly stacked packings, gas-solid trickle flow over ... 11

## S
silane decomposition reaction ..................... 77
silicon powder elutriation, prediction of ............... 77
simulated incinerator flue gas, acidic
    gas removal from ............................ 94
system parameters, the effect of ................... 83

## T
tensile strength, effect of flow conditioners on ......... 44
theoretical model, for the hydrodynamics of gas-solid ... 11
transition, from bubbling to turbulent fluidization ....... 37
turbulent fluidization, pressure on transition from ....... 37

## U
ultra-rapid
    catalytic cracking, ultrapyrolysis of ................ 69
    fluidized reactors, selection applications of .......... 69

## Z
zone studies, acceleration of ....................... 1

# SYMPOSIUM SERIES

## ADSORPTION

| | | |
|---|---|---|
| 96 Developments in Physical Adsorption | 230 Adsorption and Ion Exchange—'83 | 242 Adsorption and Ion Exchange: Recent Developments |
| 117 Adsorption Technology | 233 Adsorption and Ion Exchange—Progress and Future Prospects | 264 Adsorption and Ion Exchange: Fundaments and Applications |
| 219 Recent Advances in Adsorption and Ion Exchange | | |

## AEROSPACE

33 Rocket and Missile Technology
52 Chemical Engineering Techniques in Aerospace

## BIOENGINEERING

| | | |
|---|---|---|
| 69 Bioengineering and Food Processing | 108 Food and Bioengineering—Fundamental and Industrial Aspects | 172 Food, pharmaceutical and bioengineering—1976/77 |
| 84 The Artificial Kidney | 114 Advances in Bioengineering | 181 Biochemical Engineering Renewable Sources of Energy and Chemical Foodstocks |
| 86 Bioengineering ... Food | 163 Water Removal Processes: Drying and Concentration of Foods and Other Materials | |
| 93 Engineering of Unconventional Protein Production | | |
| 99 Mass Transfer in Biological Systems | | |

## CRYOGENICS

224 Cryogenic Processes and Equipment 1982
251 Cryogenic Properties, Processes and Applications 1986

## CRYSTALLIZATION

| | | |
|---|---|---|
| 110 Factors Influencing Size Distribution | 215 Nucleation, Growth and Impurity Effects in Crystallization Process Engineering | 253 Fundametnal Aspects of Crystallization and Precipitation Processes |
| 193 Design Control and Analysis of Crystallization Processes | 240 Advances in Crystallization From Solutions | |

## DRAG REDUCTION

11 Drag Reduction
130 Drag Reduction in Polymer Solutions

## ENERGY

### Conversion and Transfer

| | | |
|---|---|---|
| 5 Heat Transfer, Atlantic City | 119 Commercial Power Generation | 216 Processing of Energy and Metallic Minerals |
| 57 Heat Transfer, Boston | 138 Heat Transfer—Research and Design | 225 Heat Transfer—Seattle 1983 |
| 59 Heat Transfer, Cleveland | 162 Energy and Resource Recover from Industrial and Municipal Solid Wastes | 236 Heat Transfer—Niagra Falls 1984 |
| 75 Energy Conversion Systems | 174 Heat Transfer, Research and Application | 245 Heat Transfer—Denver 1985 |
| 79 Heat Transfer with Phase Change | 189 Heat Transfer—San Diego 1979 | 257 Heat Transfer—Pittsburgh 1987 |
| 87 Advances in Cryogenic Heat Transfer | 202 Transport with Chemical Reactions | 263 Heat Transfer—Houston 1988 |
| 113 Convective and Interfacial Heat Transfer | 208 Heat Transfer—Milwaukee 1981 | 269 Heat Transfer—Philadelphia 1989 |
| 118 Heat Transfer—Tulsa | | |

### Nuclear Engineering

| | | |
|---|---|---|
| 53 Part XIII | 106 Part XXII | 169 Developments in Uranium Enrichment |
| 56 Part XIV | 119 Commercial Power Generation | 191 Nuclear Engineering Questions Power Reprocessing, Waste, Decontamination Fusion |
| 95 Part XX | 168 Heat Transfer in Thermonuclear Power Systems | 221 Recent Developments in Uranium Enrichment |
| 104 Part XXI | | |

## ENVIRONMENT

| | | |
|---|---|---|
| 78 Water Reuse | 165 Dispersion and Control of Atmospheric Emissions, New-Energy-Source Pollution Potential | 201 Emission Control from Stationary Power Sources: Technical, Economic and Environmental Assessments |
| 97 Water—1969 | 170 Intermaterials Competition in the Management of Shrinking Resources | 207 The Use and Processing of Renewable Resources—Chemical Engineering Challenge of the Future |
| 115 Important Chemical Reactions in Air Pollution Control | 171 What the Filterman Needs to Know About Filtration | 209 Water—1980 |
| 122 Chemical Engineering Applications of Solid Waste Treatment | 175 Control and Dispersion of Air Pollutants: Emphasis on $NO_X$ and Particulate Emissions | 210 Fundamentals and Applications of Solar Energy II |
| 124 Water—1971 | 177 Energy and Environmental Concerns in the Forest Products Industry | 211 Research Trends in Air Pollution Control: Scrubbing, Hot Gas Clean-up, Sampling and Analysis |
| 126 Air Pollution and its Control | 184 Advances in the Utilization and Processing of Forest Products | 213 Three Mile Island Cleanup |
| 133 Forest Products and the Environment | 188 Control of Emissions from Stationary Combustion Sources Pollutant Detection and Behavior in the Atmosphere | 223 Advances in Production of Forest Products |
| 137 Recent Advances in Air Pollution Control | 195 The Role of Chemical Engineering in Utilizing the Nation's Forest Resources | 232 Applications of Chemical Engineering in the Forest Products Industry |
| 139 Advances In Processing and Utilization of Forest Products | 196 Implications of the Clean Air Amendments of 1977 and of Energy Considerations for Air Pollution Control | 239 The Impact of Energy and Environmental Concerns on Chemical Engineering in the Forest Products Industry |
| 144 Water—1974: I. Industrial Wastewater Treatment | 198 Fundamentals and Applications of Solar Energy | 243 Separation of Heavy Metals and Other Trace Contaminants |
| 145 Water—1974: II. Municipal Wastewater Treatment | 200 New Process Alternatives in the Forest Products Industries | 246 Advances in Process Analysis and Development in the Forest Products Industries. |
| 146 Forest Product Residuals | | 265 Resource Recovery of Municipal Solid Wastes |
| 147 Air: I. Pollution Control and Clean Energy | | |
| 148 Air: II. Control of $NO_{XX}$ and $SO_X$ Emissions | | |
| 149 Trace Contaminants in the Environment | | |
| 151 Water—1975 | | |
| 156 Air Pollution Control and Clean Energy | | |
| 157 New Horizons for the Chemical Engineer in Pulp and Paper Technology | | |

## FLUIDIZATION

| | | |
|---|---|---|
| 101 Fundamental Processes in Fluidized Beds | 205 Recent Advances in Fluidization and Fluid-Particle Systems | 255 New Developments in Fluidization and Fluid-Particle Systems |
| 105 Fluidization Fundamentals and Application | 234 Fluidization and Fluid Particle Systems: Theories and Applications | 262 Fluidization Engineering: Fundamentals and Applications |
| 116 Fluidization: Fundamental Studies Solid-Fluid Reactions, and Applications | 241 Fluidization and Fluid Particle Systems: Recent Advances | 270 Fluidization and Fluid Particle Systems: Fundamentals and Application |
| 176 Fluidization Application to Coal Conversion Processes | | |

## HISTORY OF CHEMICAL ENGINEERING

100 The History of Penicillin Production
235 Diamond Jubilee Historical/Review Volume

## ION EXCHANGE

79 Adsorption and Ion Exchange Separations
219 Recent Advances in Adsorption and Ion Exchange
230 Adsorption and Ion Exchange—'83
233 Adsorption and Ion Exchange—Progress and Future Prospects
259 Recent Progress in Adsorption and Ion Exchange

## KINETICS

25 Reaction Kinetics and Unit Operations
73 Kinetics and Catalysis

## MINERALS

15 Mineral Engineering Techniques
85 Fossil Hydrocarbon and Mineral Processing
173 Fundamental Aspects of Hydrometallurgical Processes
180 Spinning Wire from Molten Metals
216 Processing of Energy and Metallic Minerals

## PETROCHEMICALS

49 Polymer Processing
127 Declining Domestic Reserves—Effect on Petroleum and Petrochemical Industry
135 The Petroleum/Petrochemical Industry and the Ecological Challenge
142 Optimum Use of World Petroleum
212 Interfacial Phenomena in Enhanced Oil Recovery

## PETROLEUM PROCESSING

103 $C_4$ Hydrocarbon Production and Distribution
127 Declining Domestic Reserves—Effect on Petroleum and Petrochemical Industry
135 The Petroleum/Petrochemical Industry and the Ecological Challenge
155 Oil Shale and Tar Sands
226 Underground Coal Gasification: The State of the Art

## PHASE EQUILIBRIA

2 Pittsburgh and Houston
6 Collected Research Papers
88 Phase Equilibria and Gas Mixtures Properties

## PROCESS DYNAMICS

36 Process Dynamics and Control
46 Process Systems Engineering
55 Process Control and Applied Mathematics
214 Selected Topics on Computer-Aided Process Design and Analysis
267 Process Sensing and Diagnostics

## SEPARATION

120 Recent Advances in Separation Techniques
192 Recent Advances in Separation Techniques—II
250 Recent Advances in Separation Techniques—III

## SONICS

109 Sonochemical Engineering

## MISCELLANEOUS

48 Chemical Engineering Reviews
70 Small-Scale Equipment for Chemical Engineering Laboratories
112 Engineering, Chemistry, and Use of Plasma Reactors
125 Vacuum Technology at Low Temperatures
143 Standardization of Catalyst Test Methods
182 Biorheology
183 The Modern Undergraduate Laboratory Innovative Techniques
185 Electro Organic Synthesis Technology
186 Plasma Chemical Processing
187 Chronic Replacement of Kidney Function
194 Hazardous Chemical—Spills and Waterborne Transportation
203 A Review of AIChE's Design Institute for Physical Property Data (DIPPR) and Worldwide Affiliated Activities
204 Tutorial Lectures in Electrochemical Engineering and Technology
206 Controlled Release Systems
217 New Composite Materials and Technology
220 Uncertainty Analysis for Engineers
228 Problem Solving
229 Tutorial Lectures in Electrochemical Engineering and Technology—II
231 Data Base Implementation and Application
237 Awareness of Information Sources
238 New Developments in Liquid-Liquid Extractors: Selected Papers From ISEC '83
244 Experimental Results from the Design Institute for Physical Property Data. I: Phase Equilibria
247 Chemical Engineering Data Sources
248 Industrial Membrane Processes
249 Measurement of High Temperatures in Furnaces and Processes
252 Thin Liquid Film Phenomena
254 Electrochemical Engineering Applications
256 Experimental Results From the Design Institute for Physical Property Data: Phase Equilibria and Pure Component Properties
258 Fiber Optics: Processing and Applications
261 New Membrane Materials and Processes for Separation
266 Diffusion and Convection in Porous Catalysts
268 Membrane Reactor Technology

## MONOGRAPH SERIES

3 The Manufacture of Nitric Acid by the Oxidation of Ammonia—The DuPont Pressure Process by Thomas H. Chilton
4 Experiences and Experiments with Process Dynamics by Joel O. Hougen
5 Present, Past, and Future Property Estimation Techniques by Robert C. Reid
6 Catalysts and Reactors by James Wei
7 The 'Calculated' Loss-of-Coolant Accident by L.J. Ybarrondo, C.W. Solbrig, H.S. Isbin
8 Understanding and Conceiving Chemical Process by C. Judson King
9 Ecosystem Technology: Theory and Practice by Aaron J. Teller
10 Fundamentals of Fire and Explosion by Daniel R. Stull
11 Lumps, Models and Kinetics in Practice by Vern W. Weekman, Jr.
12 Lectures in Atmospheric Chemistry by John H. Seinfeld
13 Advanced Process Engineering by James R. Fair
14 Synfuels from Coal by Bernard S. Lee
15 Computer Modeling of Chemical Processes by J.D. Seader
16 "High-Tech" Materials by Sheldon Isakoff
17 Separations: New Directions for an Old Field
128 Biotechnology: Status and Perspectives